Erol Başar (Ed.)

Chaos in Brain Function

Containing Original Chapters by E. Başar and
T. H. Bullock and Topical Articles Reprinted
from the Springer Series in Brain Dynamics

With 66 Figures

Springer-Verlag Berlin Heidelberg New York
London Paris Tokyo Hong Kong

Professor Dr. EROL BAŞAR
Institute of Physiology
Medical University Lübeck
Ratzeburger Allee 160
D-2400 Lübeck 1, FRG

ISBN-13: 978-3-540-52329-1   e-ISBN-13: 978-3-642-75545-3
DOI: 10.1007/978-3-642-75545-3

Library of Congress Cataloging-in-Publication Data. Chaos in brain function/Erol Başar, ed.
p. cm. "Containing original chapters by E. Başar and T. H. Bullock and topical articles
reprinted from the Springer series in brain dynamics." Includes bibliographical references.
ISBN 0-387-52329-4 (U.S.: alk. paper) 1. Brain – Mathematical models. 2. Chaotic behavior in
systems. I. Başar, Erol. II. Bullock, Theodore Holmes. III. Springer series in brain dynamics.
QP376.C47 1990 596'.0188–dc20 90-9560

© Springer-Verlag Berlin Heidelberg 1990
Softcover reprint of the hardcover 1st edition 1990

Typesetting, printing and bookbinding: Brühlsche Universitätsdruckerei Giessen
2125/3130-543210 – Printed on acid-free paper

Preface

The analysis of deterministic chaos is currently an active field in many branches of research. Mathematically all nonlinear dynamical systems with more than two degrees of freedom can generate chaos, becoming unpredictable over a longer time scale. The brain is a nonlinear system par excellence. Accordingly, the concepts of chaotic dynamics have found, in the last five years, an important application in research on compound electrical activity of the brain. The present volume seeks to cover most of the relevant studies in the newly emerging field of chaotic attractors in the brain.

This volume is essentially a selection and reorganization of contributions from the first two volumes in the *Springer Series in Brain Dynamics*, which were based on conferences held in 1985 and 1987 in Berlin. It also includes (a) a survey of progress in the recording of evoked oscillations of the brain both at the cellular and EEG levels and (b) an agenda for research on chaotic dynamics.

Although the first publications pointing out evidence of chaotic behavior of the EEG did not appear until the beginning of 1985, the presence of the pioneering scientists in this field gave the participants at the first conference (volume 1) a strong impulse toward this field. For me, as conference organizer, having been for a long time active in nonlinear EEG research, the integration of this topic was self-evident; however, the enthusiasm of the conference participants was greater than expected.

Just two years later, there were three times as many contributions to the second volume of *Brain Dynamics*, and the analysis of chaotic attractors belonged to the important building blocks. Here the topic "integrative functions of the brain" was dealt with in terms of several multidisciplinary approaches. The response to the conference was so favorable that two additional papers were added to volume 2, and the authors were highly cooperative in making revisions in and additions to their papers. The authors seemed not to be put off by the strong review procedure but rather to be motivated by it. One of the editors of volume 2, T. H. Bullock, insisted on methodological extensions and conceptual additions on the basis of extensive interaction with the authors and postconference communication among experts.

The stimulus to publish this supplementary volume lay in the interest expressed by several persons to learn about and to start research in this new field. At present it is still not easy to glean the relevant publications, distributed over several special journals and conference volumes. For this reason, the editor and the publisher decided that a book of about 170 pages might more easily reach scientists working in this new field than the two full volumes covering a much broader spectrum of neuroscience.

Now greater than ever, we need new windows that can add to the growing list of techniques for comprehension of electromagnetic activity and local signs of change in the brain. In the "Epilogue" to *Brain Dynamics* volume 2 Bullock points out the importance of the new window as follows: "The contributions of Babloyantz, Rössler, Başar and Röschke, Skinner, Mpitsos, and others make me hopeful that we will soon see the dimensionality of many parts of the brain at the same time, second by second, in cats, catfish, and octopus, as rest changes into arousal, directed attention and recognition." In fact, shortly thereafter it became possible to realize some of these wishes. It is my hope that this book as well will prove to address the genuine research questions of the future.

The editor wishes to express his sincere thanks to Professor T. H. Bullock for providing his constructive critique on the preliminary survey and for adding his agenda for research on chaotic dynamics to the present volume.

EROL BAŞAR

Contents

List of Contributors

BABLOYANTZ, A., Department of Physical Chemistry II,
Free University of Brussels, P.O. Box 231, Boulevard du Triomphe,
B-1050 Brussels, Belgium

BAK, C. K., Physics Laboratory I, Technical University of Denmark,
DK-2800 Lyngby, Denmark

BAŞAR, E., Institute of Physiology, Medical University Lübeck,
Ratzeburger Allee 160, D-2400 Lübeck, FRG

BULLOCK, T. H., Department of Neurosciences A-001,
School of Medicine, University of California, San Diego, La Jolla,
CA 92093, USA

BURTON, W. D., Neurosciences Signal Analysis Laboratory,
Department of Psychiatry and Behavioral Science,
University of Texas Medical School, Houston, TX 77030, USA

ELBERT, T., Psychiatry, 116A3, Veterans Administration Medical Center,
3801 Miranda, Palo Alto, CA 94304, USA

FREEMAN, W. J., Department of Physiology-Anatomy,
University of California, Berkeley, CA 94720, USA

FULTON, K., Neurophysiology Section, Department of Neurology
and Neuroscience Program, Baylor College of Medicine,
Houston, TX 77030, USA

GRAF, K. E., Department of Clinical and Physiological Psychology,
University of Tübingen, Gartenstr. 29, D-7400 Tübingen, FRG

HUDSON, J. L., Department of Chemical Engineering,
University of Virginia, Charlottesville, VA 22901, USA

LANDISMAN, C. E., Neurophysiology Section, Department of Neurology
and Neuroscience Program, Baylor College of Medicine,
Houston, TX 77030, USA

LEBECH, J., Physics Laboratory I, Technical University of Denmark,
DK-2800 Lyngby, Denmark

MARTIN, J. L., Neurophysiology Section, Department of Neurology
and Neuroscience Program, Baylor College of Medicine,
Houston, TX 77030, USA

MITRA, M., Neurophysiology Section, Department of Neurology
and Neuroscience Program, Baylor College of Medicine,
Houston, TX 77030, USA

MOMMER, M. M., Neurophysiology Section, Department of Neurology
and Neuroscience Program, Baylor College of Medicine,
Houston, TX 77030, USA

MPITSOS, G. J., M.O. Hatfield Marine Science Center,
Oregon State University, Newport, OR 97365, USA

RÖSCHKE, J., Institute of Physiology, Medical University Lübeck,
Ratzeburger Allee 160, D-2400 Lübeck, FRG

RÖSSLER, O. E., Institute of Physical and Theoretical Chemistry,
University of Tübingen, Auf der Morgenstelle 8,
D-7400 Tübingen, FRG

SABERS, A., Department of Neurology, Hvidovre Hospital,
DK-2650 Hvidovre, Denmark

SAERMARK, K., Physics Laboratory I, Technical University
of Denmark, DK-2800 Lyngby, Denmark

SALTZBERG, B., Neuroscience Signal Analysis Laboratory,
Department of Psychiatry and Behavioral Science,
University of Texas Medical School, Houston, TX 77030, USA

SKINNER, J. E., Neurophysiology Section, Department of Neurology
and Neuroscience Program, Baylor College of Medicine,
Houston, TX 77030, USA

Chaotic Dynamics and Resonance Phenomena in Brain Function: Progress, Perspectives, and Thoughts

E. Başar

1 Preliminary Remarks

My main goal by writing this survey is to try to bridge our knowledge in chaos research with analysis of brain oscillatory phenomena both at the neuronal and the EEG level.

In the years following the first measurement of human EEG by Hans Berger and important developments by Lord Adrian and later Grey Walter, the pure EEG research remained somewhat in the shadow of new discoveries based on single-neuron recordings. From the beginning of 1960s the use of signal averagers enabled EEG research scientists to extract the evoked potentials from the so-called random-noise EEG. In this context, the event-related potentials that contributed highly to the understanding of cognitive functions and to clinical diagnostics were considered as deterministic signals whereas EEG was considered mostly as pure noise.

In 1980 I stated strongly that we must not always consider the spontaneous oscillations in brain electrical activity as a background noise. On the contrary, in my own approach on field potentials, I assumed that the EEG must be considered as one of the most important oscillations affecting the production and conduction of signaling in the brain (Başar 1980). This view was based on experiments with compound potentials from the cat and human brain. In this approach some EEG fragments were considered as internal evoked potentials coming from yet unknown or hidden sources. Further, evoked potentials were considered as forced (or evoked) oscillations following known and deterministic inputs. Later, I tentatively assumed that the EEG has a strange attractor (Başar 1983a, b). According to these earlier steps and thoughts in considering the EEG as a quasideterministic signal, I brought experts together at conferences in 1983, 1985, and 1987.

The most important development in the field of chaotic dynamics is to me, without any question, the discovery by Babloyantz et al. (1985) who showed the strange attractor behavior of EEG during the slow-wave sleep (SWS) stage. In the same year we were able to show similar results from intracranial structures of the cat brain; Rapp and coworkers published results on chaotic dynamics in the same year (Röschke and Başar 1985; Rapp et al. 1985a, b).

With these introductory remarks I emphasize that the slogan "The EEG is not noise, but is a quasideterministic signal" came from a need for interpreting results of experiments by many neurophysiologists. The new development in research of "chaos in brain function" is fascinating. However, this field cannot be considered as an isolated research field. In the period of 1985–1989 during which chaotic EEG results were substantially developed other noteworthy progress in the study of oscillatory phenomena and neural network resonance at the cellular level was also achieved. Important papers from three groups working on resonance phenomena of the brain attracted the attention of the neurophysiological community (the groups of Singer, of Llinás, and of Eckhorn). Therefore, I will also treat here recent progress on oscillatory and resonance phenomena parallel to chaotic dynamics.

2 What Is Chaotic Dynamics About?

2.1 The New Interdisciplinary Approach

The newly introduced conceptual change in brain dynamics can be a useful one in our efforts to understand the brain, even considering preliminary difficulties, shortcomings, and useful criticisms. It is part of the rationale of this chapter to help the reader with an appropriate orientation for reading this book. Over the past decade chaos has become a very lively project of scientific study. In the present context we understand chaos to refer to irregular fluctuation which is, however, described by deterministic equations, as distinct from indeterminate fluctuation that obeys the definitions of randomness (see Fig. 1).

Thus, chaos introduces an intermediate between strict determinism and randomness. A truly deterministic description of chaotic dynamics requires infinite precision in the choice of initial conditions and, thus, is a scientific chimera. Based on this evidence, Schuster (1988) assumes that chaos introduces a fundamental uncertainty which is more general than Heisenberg's uncertainty in quantum mechanics.

The problem of nonlinear dynamics originates in planetary motions. Henri Poincaré was the first to investigate the complex behavior of simple mathematical systems. He analyzed topological structures in phase space and discovered that the equation for the motion of planets could display an irregular or chaotic motion (Poincaré 1892). A mathematical basis for this behavior was later given by Birkhoff (1932). However, only in 1963 in a model of boundary layer convection did Lorenz discover that a system of three first-order nonlinear differential equations can exhibit a chaotic behavior. Contrary to Poincaré's example Lorentz discovered deterministic chaos in dissipative systems.

The differential equations of the Lorenz attractor are:

$$\dot{x} = \sigma(y - x),$$
$$\dot{y} = x(r - z) - y,$$
$$\dot{z} = xy - bz$$

where σ, b are constant parameters.

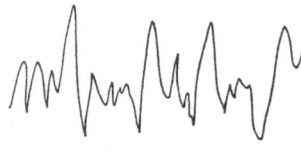

$$\longrightarrow$$
time

Fig. 1. Irregular motion

Table 1. Partial list of systems found to be chaotic (From Schuster 1988)

Forced pendulum
Fluids near the onset of turbulence
Lasers
Nonlinear optical devices
Josephson junctions
Chemical reactions
Classical many-body systems (three-body problem)
Particle accelerators
Plasmas with interacting nonlinear waves
Biological models for population dynamics
Stimulated heart cells

It is understood that some examples of each class have been studied, and that no statement can be made as to the generality of chaos in each class.

Although this model was derived for the convection instability in fluid dynamics, the single-mode laser is also described by equations equivalent to the Lorenz equations (Haken 1983).

The essential results derived from the Lorenz equations were the following

1. Oscillations with a pseudorandom time behavior (or chaotic behavior).
2. Trajectories which oscillate chaotically for a long time before they run into a static or periodic stable stationary state (preturbulence).
3. Some trajectories alternate between chaotic and stable periodic oscillations (intermittency).
4. For certain parameter values trajectories appear chaotic although they stay in the neighborhood of an unstable periodic oscillation (noisy periodicity; see Kaiser 1988).

The phenomenon of deterministic chaos as clearly described by Lorenz' system is abundant in nature and technical systems and has important functional consequences. In Table 1 some nonlinear systems that display deterministic chaos are presented. The list is far from being complete, but it gives a good idea how different functions such systems can have.

As Schuster (1988) states, the chaotic behavior in time is due neither to external sources of noise (there are none in Lorenz' equations) nor to an infinite number of degrees of freedom (in the Lorenz system there are only three degrees of freedom). Nor are these systems associated with quantum mechanics; the systems considered are purely classical.

2.2 Some Preliminary Remarks on the Nonlinear Approach to EEG and Brain Function

In a recently published popular book on chaos (Gleick 1987) the advocate of the new science goes so far as to say that "twentieth-century science will be remembered for just three things: *relativity, quantum mechanisms*, and *chaos.*" In other words, chaos has become the century's third great revolution in the physical sciences. Like the first two revolutions, chaos cuts away at the tenets of Newtonian physics. Can this development also be useful and of such great importance in brain research? For a new evaluation of EEG, event-related rhythms, or sensory evoked rhythmicities a renaissance based on this approach is possible if neuroscientists dealing with the new concept can design new types of experiments. They must also cooperate closely with theoretical physicists who should make efforts to adjust their algorithms to evaluate shorter segments of biological activity and to avoid nonstationarity, especially for brain waves.

The conceptual renaissance with the slogan "EEG is not simple noise" is already a gigantic step or watershed. Just at this point it seems useful to mention a paradoxical aspect in chaotic neuroscience: the neuroscientists have not yet discovered how to apply the Newtonian approach to physics or quantum statistics to brain science. This difficulty has already been mentioned by Rosen (1969) and Başar (1980) and many other authors. The usefulness of predictability and of mathematical description of brain behavior has not been in the mainstream of brain research, with a few exceptions. Therefore, the big bang of applying chaotic dynamics to brain activity has struck brain scientists in a still unprepared and inappropriate stage. In this short survey we will also mention some criticisms and words of caution among seemingly useful new trends and speculations explained in the following chapters. The enthusiasm over this new branch of brain research can lead to useful results only if a proper design of experiments can be established.

3   Definitions and New Types of Expressions

For the neuroscientist who is not familiar with the jargon some explanations will be given here. This step is useful although almost all the descriptions given are contained in various chapters that follow (see Babloyantz 1988, and this volume; Röschke and Başar 1988, 1989, and this volume). In this way a preliminary global insight may be obtained.

For the nonspecialist in chaotic dynamics the reading of appropriate books is essential to obtain a deeper understanding (for example, Schuster 1988; Abraham and Shaw 1983; Babloyantz 1986; Haken 1976). However, we give in the following some important definitions to orient the reader.

Attractor. The property of a dynamic system which is manifested by the tendency under various but delimited conditions to go to a reproducible active state and stay there. The trajectory is a mathematical description of the sequence of values taken by a state variable in going from an initial or starting condition to an attractor or through a sequence of attractors (Abraham and Shaw 1983). Transition from one attractor to another is called a state change or bifurcation. Attractors can be periodic, quasiperiodic or chaotic; the last are called strange attractors. (For a number of definitions adapted to physiological systems see Freeman and Skarda 1985.)

Fixed Point. Simplest stable state solution. With increasing time, all trajectories tend to terminate in this point. Stable fix points are static attractors (see Fig. 2A).

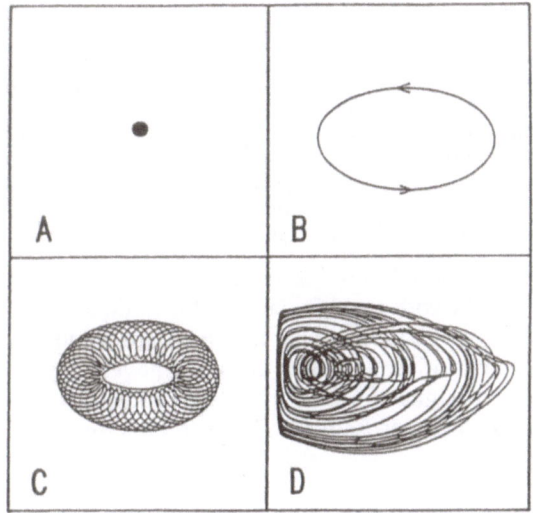

Fig. 2 A–D. Types of attractors.
(A) Fixed point. (B) Limit cycle.
(C) Torus. (D) Projection of a
strange attractor

A standard example is a pendulum that has come to rest after some time of oscillation due to friction.

Limit Cycle. Closed and recurrent trajectory in phase space. All trajectories tend to terminate in this cycle; no other closed cycle lies in its neighborhood. Without external drive, the limit cycle corresponds to a periodic stable position of the non-linear system, whose amplitude and frequency are determined by internal parameters of self-sustained oscillations. Stable limit cycles act as periodic attractors (see Fig. 2 B). The standard example is the attractor of a van der Pol oscillator. Limit cycles regularly occur with driven oscillators (Başar 1976, 1980).

Torus. The system's trajectories move on a two-dimensional toroidal surface. Two frequencies are present, oscillations around the torus and along the torus (oscillations with two incommensurable frequencies). The trajectory never closes or covers the whole torus (see Fig. 2 C). The trajectory on the torus is a quasiperiodic motion.

Strange Attractor. The manifestation of a strange attractor is its activity which appears to be random, but which is deterministic and reproducible if the input and initial conditions can be replicated (for example, Lorenz attractor or Rössler attractor; Fig. 2 D). Since they cannot in practice be replicated, the manifestation is usually that after many trajectories, the phase plane is not evenly filled, as it would be for a random time series, but is occupied by a quasipatterned line, never exactly repeated but clearly constrained.

Noise. Bullock (1976) describes the noise in the general neurophysiological approach as follows: "unwanted action" that interferes with desired signals. This author further states:

"We should recognize the sharp difference between this dictionary usage and another current usage that refers to a stochastic sequence ('whiteness'). In the first meaning, noise is determined by the state of the receiver (sleep, attention) and depends on the usefulness, regardless of the character; any unwanted sequence is regarded as noise whether it is a hiss, a whistle, or a voice. In the second meaning, noise is determined by the state of the sender (filter settings) and depends on the statistical character regardless of the use; any quasirandom sequence is regarded as noise whether it is unwanted interference or a high resolution signal. The first meaning overlooks the difficulty of knowing what may be of value to a receiver; the second overlooks the difficulty of avoiding the common English sense, as in "signal-to-noise ratio.""

According to the view of Bullock, we should not use the term "noise" unless we are prepared to claim that we know the codes and functions of the system and can recognize its signals. In the language of chaotic dynamics the noise could be defined as "a signal showing irregular motion and that does not have a finite dimension," in other words, an irregular signal whose correlation dimension D_2 does not show saturation (see Röschke 1986; Röschke and Başar 1989 and this volume).

Correlation Dimension. This has become the most widely used measure to describe chaotic behavior. A valuable first step in the study of dynamical behavior, particularly when chaos is present, is measuring its dimension and investigating how the dimensionality can change under different operational circumstances (Mees et al. 1987). A rigorous review of dimensions is given in several papers (Grassberger and Procaccia 1983; Babloyantz, this volume; Röschke and Başar 1989 and this volume). Less rigorously stated, it can be that the correlation dimension of a system's behavior is the minimum number of dimensions of a space that can contain the trajectories generated by the system. As Rapp et al. (1985 a, b) express it, the dimension of a system is its number of degrees of freedom. This definition is restricted but simple and useful. It is important to compare systems only by referring to the same quantity, usually the correlation dimension (D_2). A system is periodic if its D_2 is a whole number (e.g., 2.0, 3.0, 4.0) and chaotic if D_2 is "fractal" (e.g., 2.1, 3.9, 4.5).

4   The EEG Has a Strange Attractor: The EEG Is Not Noise

4.1   Beginning of the Development with the Slogan "EEG Has a Strange Attractor" (or EEG Reflects Deterministic Chaos)

The new trend in brain research could be initiated in the proper manner by evaluation of the correlation dimension D_2 to the brain's EEG during SWS by Babloyantz et al. (1985) and shortly after that by application of the same algorithm to some pathological cases. Following the most important pioneering work by Babloyantz and coworkers, Röschke and Başar (1985) published results on the strange attractors in several intracranial structures of the cat brain during SWS and confirmed in a general way the results of Babloyantz et al. Further, Layne et al. (1986) and Rapp et al. (1985a, b) interpreted the waking EEG as chaotic behavior.

In 1983 we also mentioned the possibility of comparing the filtered EEG with components of Navier-Stokes equations and tentatively used the expression alpha attractor, which is a type of strange attractor. Başar (1983 a, b) described the necessity of evaluating the EEG signals with nonlinear tools.

4.2   Fundamental Questions in Chaotic Dynamics of Brain Potentials

Most of the papers included in this volume show evidence of a chaotic attractor in the EEG. Further, the authors try to answer the following fundamental questions:

1. Is it possible to identify attractors for various states of brain activity? Can the activity of neural populations be described by deterministic dynamics (Babloyantz 1988, this volume; Röschke and Başar 1988, 1989 and this volume)?
2. If attractors do exist in brain activity, what are their dimensionalities? Is it possible to classify attractors as periodic, quasiperiodic, or chaotic (Babloyantz, this volume; Saermark et al. 1989)?
3. Can the transitions of brain activity be described in terms of nonlinear dynamics (Başar et al. 1988; Röschke and Başar 1989; Babloyantz, this volume; Skinner et al. 1989; Lopes da Silva et al. 1990)?
4. What is the role of chaotic dynamics in brain function (Mpitsos 1989; Röschke and Başar 1988)?
5. Can the correlation dimension characterize pathological changes (Babloyantz, this volume; Saermark et al. 1989; Graf and Elbert 1989)?

4.3   A Limited State of the Art

In Tables 2 and 3 the values of correlation dimensions (D_2) computed by several research groups under different experimental conditions are outlined together. The parameters are also included in the tables on human EEG. In Table 2 the results of measurements on human subjects are presented; in Table 3 the experiments with intracranial recording of cat brain, rat brain, and rabbit brain are shown. We will survey these results in this sections in a more extended manner.

In this newly emerging field it is impossible to perform a complete review of all articles written since a great number of new findings have not yet been completed or written. Some of the important articles are available only as confidential preprints or as reviews in preparation. Therefore, I will try to cover mostly the papers contained in this volume plus new works which have recently reached my hands and where the authors permit me to cite them. I cannot do justice to all the scientists working in this field. The field is new, the papers often have divergent values for D_2. Some workers (Rapp et al. 1989) believe the Grassberger and Procaccia (1983) algorithm generally used for the computation of D_2 must be modified for biological data and are about to announce their new algorithm. I also highly recommend the reading of the new survey by Rapp and coworkers

Table 2. Human EEG/MEG data

Reference	Parameters	Results
Babloyantz et al. (1985)	$\Delta t = 10$ ms $N = 4000$ $\tau = 20$ ms EEG	Sleep stage 2: $\quad D_2 = 5.03$ Sleep stage 4: $\quad D_2 = 4.0\text{--}4.4$ Awake, alpha activity: $\quad D_2 = 6.1$ Beta: $\quad D_2 -$ no saturation
Rapp et al. (1986)	$\Delta t = 2$ ms $N = 1000\text{--}4000$ $\tau = 10\text{--}20$ ms EEG	Eyes closed, relaxed: $\quad D_2 = 2.4$ (N = 1000) $\quad D_2 = 2.6$ (N = 4000) Eyes closed, counting: $\quad D_2 = 3.0$ (N = 4000)
Layne et al. (1986)	$\Delta t = 2$ ms $N = 1000\text{--}15\,000$ $\tau = 20$ ms (occipital) $\tau = 40$ ms (vertex) EEG	Awake, occipital: $\quad D_2 = 5.5\text{--}6.6$ Awake, vertex: $\quad D_2 = 6.5\text{--}7.7$
Başar et al. (1989 b; 1990; this volume)	$\Delta t = 30$ ms $N = 16\,384$ points $\quad$(segments of 3 min) sampling frequency: $\quad \Delta f = 100$ Hz EEG	Eyes closed, occipital/vertex/ parietal/frontal: $\quad D_2 = 5.5\text{--}8$ (Finite dimension only when data prefiltered between 5 and 15 Hz)
Dvorak and Siska (1986)	$\Delta t = 5$ ms $N = 1000\text{--}12\,000$ $\tau = 40$ ms EEG	Eyes closed: $\quad D_2 = 3.8\text{--}5.4$ (N = 1000) $\quad D_2 = 8\text{--}10$ (N = 12\,000)
Van Erp et al. (1987)	$\Delta t = 5\text{--}10$ ms $N = 1000\text{--}10\,000$ $\tau = 15\text{--}75$ ms EEG	Alpha rhythm: $\quad D_2 = 5\text{--}6$ (N = 1000) $\quad D_2 = 7\text{--}8$ (N = 10\,000) Beta rhythm: $\quad D_2 -$ no saturation
Babloyantz et al. (1986, this volume)	$\Delta t = 0.83$ ms $N = 6000$ $\tau = 16\text{--}60$ ms EEG	Creutzfeldt-Jakob disease: $\quad D_2 = 3.7\text{--}5.4$ Epileptic attack: $\quad D_2 = 2.05$
Saermark et al. (1989; personal communication)	$\Delta t = 10$ ms $N = 4000\text{--}8000$ $\tau = 100$ ms MEG	Healthy subject: $\quad D_2 = 11$ Epilepsy (2 patients): $\quad D_2 = 7$ Epilepsy (2 patients): $\quad D_2 -$ no saturation

D_2, Correlation dimension; N, number of data points; τ, time shift; Δt, sampling time; SWS, slow-wave sleep stage; REM, rapid-eye movement sleep; MEG, Magnetoencephalography.

Table 3. Intracranial EEG (animal experiments)

Reference	Results
Başar et al. (1988), Röschke and Başar (1985, 1988)	Cat, SWS, cortex (epidural): $D_2 = 5.0 \pm 0.1$ Cat, SWS, hippocampus: $D_2 = 4.0 \pm 0.07$ Cat, SWS, reticular formation (mesencephalon): $D_2 = 4.4 \pm 0.07$ (the most stable data)
Röschke and Başar (1989)	Cat, inferior colliculus: $D_2 = 6.7$ Cat, reticular formation (mesencephalon): $D_2 = 7.05$ (unstable attractor, waking state, attractor properties in only 25% of recording time, „high frequency attractor,“ data filtered between 100 and 1000 Hz
Röschke and Başar (1989)	Cat, waking state, hippocampus: $D_2 = 4.00$ (during synchronized hippocampal theta activity)
Lopes da Silva et al. (1990)	Rat, hippocampus: $D_2 = 2–3$ or higher (unstable depending on location and on existence of epileptic discharge)
Skinner et al. (1989; this volume)	Rabbit, olfactory bulb: $D_2 = 5–6$ (event-related shifting from 5 to 6 in evoked activities with odor targets)

(1989), which contains substantial new findings and also treats the important window with event-related potentials.

4.4 Some Remarks on Behavioral Experiments with Animals

In this survey the important precautions strongly suggested by Babloyantz, Rössler, and Röschke and Başar concerning the length of recording and the frequency limitation will limit our consideration of papers departing from the suggested rules. I refer to the experimental work of Skinner et al. (1989; and this volume) as an important conceptual experiment. As also stated by Röschke and Başar (1989 and this volume), experiments designed to measure transitions from one stage to another are pertinent to the analysis of chaos in brain function even if it is possible to show only the relative changes in D_2, where the absolute value may be debated.

With these reservations in mind the claims of Skinner et al. (1989) deserve examination. One of the most important aspects of the data by Skinner et al. is that the same odor can also evoke an *increase* in the dimensionality of an attractor if it is novel rather than familiar. In other words, the dimensionality changes if the attractor is event related.

Furthermore, Skinner et al. mention the brain-heart connection and assume that their findings are of medical importance since they observe that the neural activities in the frontocortical brain stem are both necessary and sufficient to trigger lethal cardiac arrhythmias. Resting on results by Skinner and Reed (1981), they further state that:

1. Descending activity in the cardiac nerve is necessary for the occurrence of the lethal, ventricular fibrillation.
2. Combined stimulation of sympathetic activity produces the most malignant type of descending neural activity.
3. The event-related increase in the dimensionality of an attractor in the brain may be projected to the heart (frontocortical brain stem pathway).

Although the hypotheses of Skinner et al. (1989) seem to be highly pretentious, their idea is an interesting one since some physiological data might support "resonant frequency" in the brain stem–cardiovascular connections. In several papers Gebber et al. (Gebber and McCall 1976; Gebber 1980; Gebber and Barman 1980, 1981) showed that sympathetic and parasympathetic ganglia of the heart and of the kidney depicted a type of electrical activity in the EEG frequency range. Accordingly, new designs considering the brain attractors and possible cardiovascular attractors could in future enrich our knowledge about the integrating function of the CNS. The measurements of Gebber and coworkers must therefore be studied carefully.

Recently the group of Keidel and Pöppl has published a number of studies on possible central oscillators synchronizing brain waves and muscle oscillations in man (Keidel et al. 1987, 1989). Furthermore, these groups report the large amount of data on minute rhythm in human physiology. On the basis of their observations Keidel and coworkers postulate a temporally well-ordered and possibly oscillating functional coupling between different spinal and supraspinal parts of the CNS. They support the idea of a central "clock" synchronizing supraspinal neurons and motoneurons with a clear rhythmicity. In their work there is again a strong trend towards "a common brainstem system." The scope on integrative function of CNS is highly enriched by studies of this group.

The problem of nonstationarity is also taken into consideration by Mpitsos (1989). The historical perspective of this paper is relevant, and it should be read by everyone working in this field. The consideration of the brain as a noisy processor as described by Adey (1972) is also explained by Mpitsos: Is it possible to envisage an information-processing system in which the very presence of an ongoing noiselike activity produced no degradation of information-handling ability, and might even enhance it? About noise and uncertainty at the neuronal level I further recommend the reading of the review by Bullock (1976).

One of the most fundamental aspects of the chapter by Mpitsos (this volume) is his interpretation of "why chaos is important." He assumes that chaos may have a role in the generation of rapid adaptation to changing environments. Nonvarying signals, such as limit cycles, carry no new information into the future other than what they contain at some previous time. I strongly support this view since experiments of Başar (1980) and Başar et al. (1983 a, b) have shown in animal and human recordings that brain responsiveness in the form of evoked potentials is highly reduced or disappears if a brain structure shows limit cycle activity. (Good examples are the regular theta activity of hippocampus during which only very small amplitude evoked potentials can be recorded.)

Mpitsos also assumes that long-term evolution of chaos is unpredictable, and that such unpredictability represents a gain of information by which the brain naturally creates new response possibilities.

4.5   Sleep Results Are the Most Stable Results

Recently Röschke et al. (1989; and personal communication) published results on D_2 during sleep stages I, II, III, IV and REM sleep. This group found $D_2 = 4.25$ for SWS stage (stage IV) and $D_2 = 6$ for REM sleep as means after computation on ten healthy subjects. As Röschke et al. (1989) state, these results are in good accordance with the findings of Babloyantz et al. (1985). It is noteworthy that the standard deviation from the mean is small during SWS stages but decreases during REM and sleep stage I. As we know from studies during waking state, the fluctuations are enormous (see Tables 2, 3; Figs. 3, 4).

The findings of Başar et al. (1988) and Röschke and Başar (1985, 1989) on the D_2 of intracranial structures of the cat brain were also in good agreement with human data. During SWS there is a complete absence of motor functions and also probably absence of conscious mental activity. Can these findings and considerations be an important starting base for physiological interpretations? I think that investigators will soon point out this important behavioral stability as revealed by D_2.

4.6   Dimensionality of Brain Waves in Pathology

The first evaluation of D_2 of epileptic activity was performed by the group of Babloyantz, who found that D_2 could reach values as low as $D_2 = 2$. Furthermore, Babloyantz (this volume) applied the algorithm of D_2 evaluation to Jakob-Creutzfeldt disease and mentioned the use of the concept for clinical classification.

Although the results of Saermark et al. (1989) and Graf and Elbert (1989) are different from those of Babloyantz, it must be remarked that all results on D_2 in epileptic patients have much lower values than in healthy subjects. Divergence in the absolute number may stem from recording techniques, the patterns of epileptic discharges, and the use of filters (see also Sect. 4.4).

It is recommended here that the evaluation of D_2 should be accompanied by other pattern recognition algorithms or spectral analysis.

As an additional reading to papers included in this volume the paper by Babloyantz and Destexhe (1986) is highly recommended since it is one of the most fundamental papers written on chaotic brain dynamics. The paper is important because of its presentation of a spectrum of the methods and the adequate use of the concept.

4.7   Hippocampal Activity, Epileptic Activity in Hippocampus

Lopes da Silva et al. (1990) studied the D_2 of EEG signals recorded during epileptic seizures in the rat brain. Their results indicated that the dimensionality of the signals varies as a function of the part of the hippocampus and of the time during the course of a kindled epileptic seizure. They further showed that it is not possible to state that EEG signals can be represented, in general terms, as generated necessarily by low-dimensional chaotic systems; however, during given states of epileptiform discharges this might be the case. One of the important statements of Lopes da Silva et al. is that during the waking state there are also "noisy states," meaning during long periods of time. At this point we would also like to mention the results of Röschke and Başar (1989 and this volume), who analyzed the chaotic behavior of hippocampal theta activity of the cat brain. The hippocampus shows "strange attractor behavior" and not "pure noise" only during defined regular theta activity.

Freeman (1989 and this volume) and Freeman and Skarda (1985) took into view that the evaluation of a number as D_2 is alone insufficient to describe the brain states; they emphasized the usefulness of understanding the brain function reflected in geometry (multidimensional phase portraits).

4.8   Some Critical Views

Rössler, who has introduced several important concepts in the chaotic hierarchy and describes systems with high dimensionality as "hyperchaotic" systems, took the view that the brain is a giant dynamic system. Rössler and Hudson (1989 and this volume) emphasized that it would not be surprising to find a "maintenance activity" of periodic or quasiperiodic type in the brain, although the possibility of a low-dimensional chaotic process in addition has yet to be demonstrated. Of course, a maintenance activity of a low-dimensional chaotic type is also a possibility which cannot be ruled out as a priority. Accordingly, these authors introduced a method which may be of help in discriminating between "hyperchaos generators," on the one hand, and low-dimensional, but closer to quasiperiodic, time sequences of a similar phenomenological dimensionality on the other hand.

However, I must mention here that the functional relevance of 10-Hz activity as discussed in Sect. 5 of this survey does not support the view that rhythms can be considered only as "maintenance activity." At least the experiments of Sect. 5 show that 10-Hz activity is often related to mental function.

One of the most important papers, written by Babloyantz (this volume), contains a critical view on the estimation of D_2. In using the algorithm of Grassberger and Procaccia (1983), the adequate choice of sampling frequency, time delay, data length, and upper limit of D_2 are of vital importance in order to avoid errors in estimation of D_2 and to judge whether the EEG data are chaotic or noisy.

Röschke and Başar (1989 and this volume) mentioned that the use of power spectra might help to elucidate discrepancies between results of various research groups, especially from recordings during the waking state, because during the same recording session large fluctuations in D_2 can be observed. They further assume that power spectra of the EEG are a brain descriptor complementary to D_2. The exclusive use of D_2 for describing the brain state can lead to misinterpretations. The crucial example is the comparison of D_2 of the cat cortex during SWS and of the cat hippocampus (for both cases $D_2 = 4$) whereas the power spectra of the EEG are completely different. In other words, without complementary use of the power spectra the interpretation of D_2 for a given brain state can lead to several inappropriate identifications (see also Figs. 3, 4).

Albano et al. (1987, 1988) also emphasize the necessary number of data points, the sampling interval, and stationarity in estimating attractor dimension. These authors state that "getting a number is easy; getting a dynamically meaningful number can be difficult." To the important critical views of Albano et al. (1987, 1988) I want to add that not only getting a dynamically meaningful number but also its adequate physiological interpretation should also be an essential goal in chaos analysis.

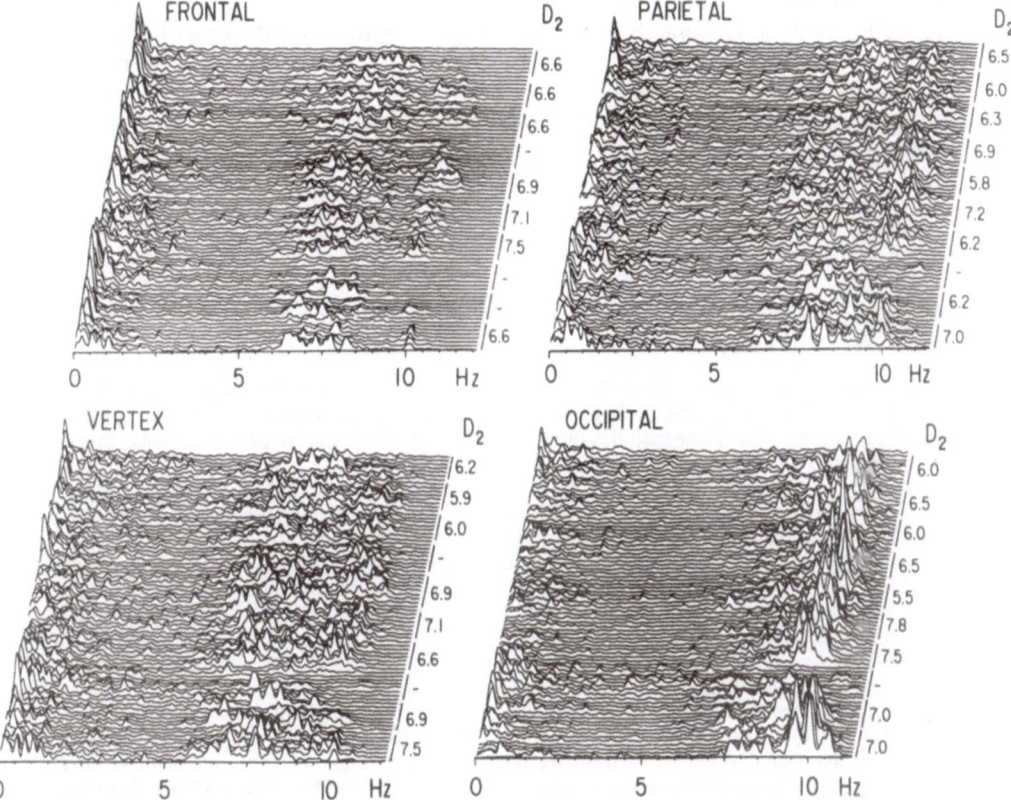

Fig. 3. A comparative presentation of power spectra (compressed spectral arrays) and the correlation dimension D_2. Each D_2 value and the adjacent spectra were calculated from the same EEG segments. Simultaneous recordings from the same subject in frontal, central, parietal and occipital locations. (From Başar et al. 1990)

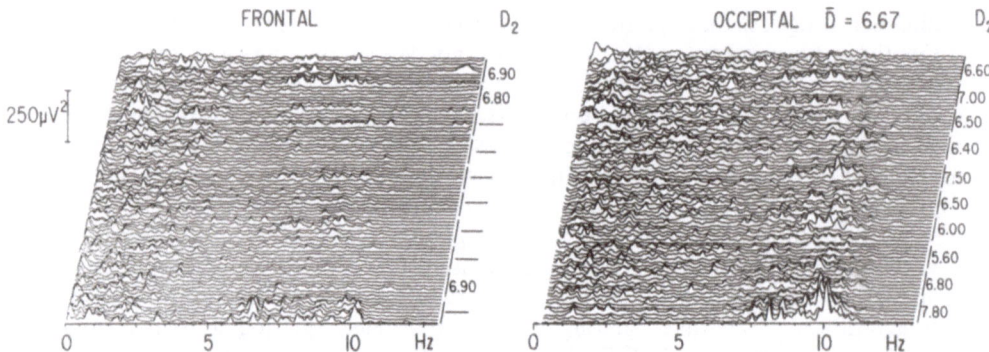

Fig. 4. A comparative presentation of power spectra and D_2 for frontal and occipital location. Another subject than that in Fig. 3 (From Başar et al. 1990)

The importance of the attractor dimension and some primary results on D_2 of EEG records are also described by Rapp et al. (1985a, b). These are among the pioneering papers treating also the physiological significance of control systems. According to these authors, a disordered system may be following a very simple, potentially discoverable, dynamical law.

I understand that the efforts of Rapp and his group also have the goal to use results obtained from chaotic dynamics to generalize the theory of physiological control systems.

4.9 Further Questions by Bullock

Bullock (1989b) has raised relevant questions concerning noise and use of D_2 which we repeat here:

For example, if the curve of correlation dimension (sometimes called D_2) against embedding dimension saturates to a clearly horizontal plateau, one is permitted to say that there is negligible "noise." The latter includes intrinsic random activity that need not be noise in the communication sense of useless interference. This is a strong statement for the electrocorticogram (Röschke and Başar, Babloyantz, Skinner et al., all this volume) or for gastropod neuron spiking (Mpitsos, this volume). We could not heretofore have said any such thing, and would have doubted any claim to this effect. As Mpitsos points out, instead of assuming, as we have, that deterministic signals ride on a background of random variation, it is now possible to show, in given cases, that the latter is insignificantly small and the observed fluctuation is essentially deterministic. It is now possible to distinguish samples quantitatively, by fractional dimensions; these may represent a minimum estimate of the nonlinear interactions as well as the number of distinct parameters like incommensurable periodicities. One might presume therefore that dimensionality grades complexity in a definable sense.

The physiologist evaluating these methods, or results reported from their use, needs to be given some feeling for their sensitivity to several factors. Röschke (1986) implies that extrinsic white noise has a small effect on the estimate of correlation dimension if the signal-to-noise ratio is > 10; it is not clear whether this is true for systems with higher as well as lower D_2. It would help to know how large must be the amplitude of some component in order to exert an effect on D_2; for example, when does an incommensurable periodicity newly appearing, and independent of the rest of the components, reach a threshold amplitude and add its contribution to D_2? How much effect does a single strong rhythm have, presumably suppressing D_2 by pushing some of the weaker signals below threshold as a proportion of the total power?

The reading of the papers by Bullock (1989 a, b) is highly recommended. These deal with a number of concepts including the synchrony of neuronal populations and a number of important concepts in brain research in addition to chaotic dynamics.

5 Brain Alpha Attractor

Since the discovery of the alpha rhythm by Hans Berger one of the biggest puzzles in electroencephalography has been the physiological understanding of their origin, their relation to sensory and cognitive functions of the brain, and not least their interactions as an indicator of the brain state.

According to Storm van Leeuwen (1977), if one understands the alpha rhythm, he will most probably understand the other EEG phenomena. Therefore, it is not surprising that after the first analysis of the electrical activity during the SWS state neuroscientists tried to measure the dimensionality of the alpha activity. The fact that several authors have worked on this problem is displayed well in the Table 2. To me it is not astonishing that there are discrepancies in the data evaluated by several investigators. Babloyantz and coworkers originally did not find any saturation during alpha activity, but they later described a D_2 of about 6 (Babloyantz, this volume). The data of some other authors vary between 3 and 8. Başar et al. (1989 b), Röschke and Başar (1989), and Başar et al. (1989 b, 1990) took another step concerning alpha activity. By filtering the EEG in a frequency range between 5 and 15 Hz during a state of high alpha activity and performing measurements on 3-min segments from 30-min records for a number of subjects, they found that under such measuring conditions the D_2 of the alpha activity fluctuated between 5 and 8 (see Table 4), even though stationarity seemed quite good over the 30 min. The simultaneously evaluated power spectra (Fig. 3) are also different. There is usually no saturation if the digital filters are not applied. On the other hand, in the measurements of the same subject there are time periods during which there is no saturation in any of the seven electrode sites, or in some of them. In some subjects one usually sees fractal dimensions between 5 and 8, but in some locations (e.g., frontal) almost never a 3-min segment with good saturation. Table 4 and Fig. 4 show large differences between the D_2 of occipital and frontal recordings.

These results help somewhat to account for the discrepancies of results among several authors – not by explaining but by confirming variance. There are large fluctuations in the dimensionality of alpha waves. This means that in this frequency range the brain has two types of behavior: noisy behavior and strange attractor behavior. Başar et al. (1989 b) concluded that the problem of the dimensionality of alpha waves or even whether the alpha activity presents a strange attractor behavior should be considered from a functional viewpoint. This means that the design of the experiment and expected results should also be taken into account for interpretation of D_2. We come here to the definition of the strange attractor in Sect. 3, in which it is stated that the strange attractor has the ability to show "reproducible patterns" if initial conditions could be kept constant.

Table 4a, b. Changes in D_2 over 30 min: values for one subject on successive 3-min samples from a 30-min EEG during good alpha, after filtering the EEG through a 5- to 15-Hz digital band-pass filter (no phase shift) (From Başar et al. 1989b)

Location	D_2 in 10 successive 3-min samples										Mean
(a) O_1	7.0	7.0	–	7.5	7.8	5.5	6.5	6.0	6.5	6.0	6.6 ± 0.7
O_2	6.2	6.6	–	6.5	7.1	6.1	6.0	6.2	7.5	7.1	6.5 ± 0.5
P_3	6.2	6.6	–	6.5	6.6	5.8	7.5	5.9	6.0	–	6.3 ± 0.5
P_4	7.0	6.2	–	6.2	7.2	5.8	6.9	6.3	6.0	6.5	6.4 ± 0.4
C_z	7.5	6.9	–	6.6	7.1	6.9	–	6.0	5.9	6.2	6.6 ± 0.5
F_3	6.6	6.6	–	6.1	7.1	5.0	–	5.5	6.0	–	6.1 ± 0.7
F_4	6.6	–	–	7.5	7.1	6.9	–	6.6	6.6	6.6	6.8 ± 0.3
(b) O_1	6.9	6.8	6.6	6.7	6.5	5.9	6.4	6.0	6.5	6.5	6.4 ± 0.3
O_2	7.1	6.9	6.8	6.5	5.9	6.1	6.2	6.1	6.3	6.6	6.4 ± 0.4
P_3	7.0	6.9	–	6.5	6.3	6.2	6.2	6.2	6.5	6.7	6.5 ± 0.3
P_4	7.5	–	–	6.5	6.0	6.5	7.1	6.2	6.7	–	6.6 ± 0.5
C_z	7.3	6.9	–	6.6	6.5	6.6	6.4	7.0	7.3	6.4	6.7 ± 0.3
F_3	6.1	6.1	6.2	5.9	6.0	5.9	5.9	6.0	6.2	6.1	6.0 ± 0.1
F_4	–	6.9	–	–	–	6.9	–	–	–	–	6.9

–, No saturation.

Figure 5 A illustrates ten sweeps of pretarget EEG activity from a subject who had the task of mentally marking a cognitive target (omitted stimuli following four regular optical repetitive stimuli with intervals of 2.5 s). Prior to target the alpha activity of the subject is phase-locked to target and shows a regular and reproducible 10-Hz pattern. Figure 5 B presents results of measurements with the same subject but with a reduction in probability of the occurrence of the target (probability decreased from 100% to 25% – every fourth to seventh stimulation randomly omitted). In this case alpha waves are no longer reproducible nor phase ordered. After a large number of experiments Başar et al. (1989 a, 1990) concluded that alpha activity has a "quasideterministic" behavior, as shown during experiments with a defined cognitive target. The alpha activity, when phase-locked to a cognitive target, fulfilled the description of the strange attractor introduced in Sect. 3. For extended information on this type of experiment with reproducible alpha pattern in phase space we refer to Başar et al. (1989 b).

In other words, the existence of a convergent dimension D_2 of filtered alpha activity was not the only prerequisite, but the reproducible alpha activity (without physical stimulation) is solid evidence that the cognitive alpha (the alpha during a state of expectation) is not simple noise. Noise cannot be generated deterministically. For a detailed explanation of experiments and results see Başar et al. (1988, 1989 a, b). At this point I come back again to the concept of noise outlined at the beginning of this survey (Sect. 3) and combine our thoughts with the "reproducible pattern." In this step I reproduce some comments of Bullock, who in additional to his work of 1976 wrote me as follows:

The arrival of synaptic potentials from many sources at the motor neuron may be a random sequence, but it is not noise in any dictionary sense. The applause at the opera may or may not be a random sequence, but it is not noise, although exactly the same sequence under other circumstances is – i.e., when it conveys no useful information and interferes with signals.

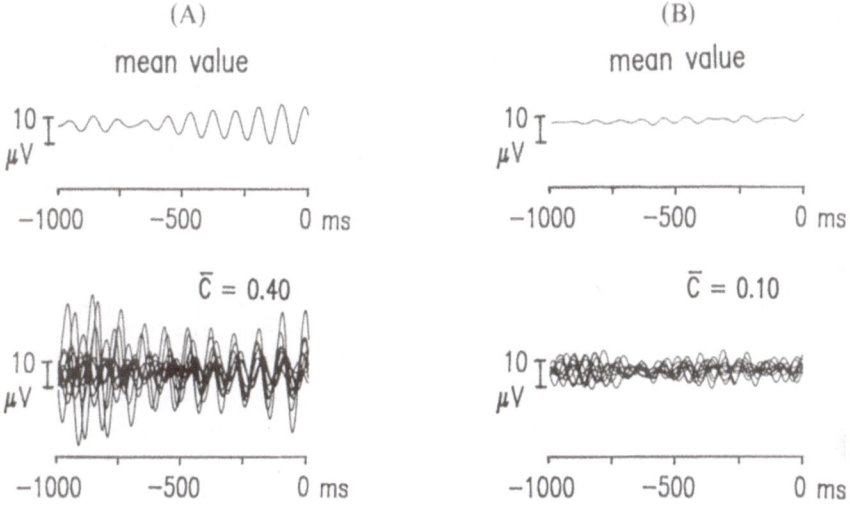

Fig. 5. (A) Pretarget EEG of a human subject during a cognitive paradigm. Subject showed repeatable 10-Hz EEG patterns prior to a target. Every fourth signal (light signal) was omitted. The time point 0 is the time of the presented target (omitted tone). The time scale from −1000 ms to 0 indicates 1-s recording time prior to the target. *Below,* ten EEG samples filtered in 8–13 Hz frequence range (digital filters without phase shift). The mean correlation coefficient of single curves is $r=0.4$ for the time period 500 ms prior to target. This means high congruency of ten EEG sweeps during the cognitive task. *Above,* mean value of ten EEG sweeps. Due to overlapping of ten EEG signals showing an almost perfect phase locking the mean EEG has large amplitude. The subject reported that his performance was good. (B) Similar paradigms due to good congruence of single sweeps as in (A) for the same subject. However, every third to seventh signal was omitted. The subject reported that he could try to do well, but he did not have a good performance, as in the experiment in (A). The correlation coefficient is low, and the mean EEG (*above*) is small in comparison to the experiment in (A). Single EEG epochs are not reproducible as in (A)

I find that the applause at the opera is a good example to come back to regarding the expressions noise and random. In fact, if alpha waves with reproducible pattern are recorded prior to or after a target event, we say that they arrive prior to or after an expected event. It is, however, possible that neural networks which produce the alpha activity can never produce alpha activity which is noise. Since in most cases we do not define or relate an activity to a target that we think is noise; but in reality there are only randomly occurring signals. At the opera we sometimes also observe rhythmic applause. Is such applause then similar to evoked or induced rhythmicities in the brain? (See also Sect. 6.)

Also to be mentioned are the important results of Freeman and Skarda (1985) with repeatable 40-Hz EEG patterns from brains of rabbits expecting specific odors. The reader here is also referred to Skarda and Freeman (1987) and open peer commentary in *Behavioral and Brain Sciences* (1987) 10:2.

Bullock (1989 a, b) stated that the contributions by Babloyantz, Rössler, Başar and Röschke, Skinner, and Mpitsos and others give him hope that we will soon see the dimensionalities in cats, catfish, and octopus of many parts of the brain simultaneously as rest changes into arousal, directed attention, and recognition. Although the hope that Bullock expressed has not yet been fully realized, the fact that recordings of successive 3-min periods in various brain sites showed large changes of D_2 between time segments and in the same segment between different locations (Table 4; Figs. 3, 4) makes this hope seem to be realizable soon. It will be now a question of time to include in evaluations the states of arousal and attention.

We also mention here once more the following statement made by Babloyantz in 1979:

> If deterministic chaos is detected from a single-channel recording, it indicates the presence of chaotic activity in the recorded site. Nothing guarantees that the time series from an adjacent lead will show a chaotic activity or if it does, whether we are dealing with the previous attractor.

It seems that one site in the brain (frontal) can show noise behavior over long periods whereas another site (occipital) can under certain conditions show chaotic behavior at the same time.

A most important publication on alpha activity is that by Pfurtscheller et al. (1988), in which the authors describe the alpha band rhythm and its event-related desynchronization. Emphasis is given to comparing the lower (6–10 Hz) and upper (9–13 Hz) alpha-bands, i.e., to the variety of rhythmicities in the frequency scale. An important remark by Walter (1964), who is known as the discoverer of the contingent negative variation, was also mentioned by Pfurtscheller et al.; we reproduce this here:

> We've managed to check the alpha band rhythm with intracerebral electrodes in the occipital-parietal cortex; in regions which are practically adjacent and almost congruent one finds a variety of alpha rhythms, some of which are blocked by opening and closing the eyes, some are not, some are driven by flicker, some are not, some respond in some way to mental activity, some do not. What one sees on the scalp is a spatial average of a large number of components, and whether you see an alpha rhythm of a particular type or not depends upon which component happens to be the most highly synchronized process over the largest superficial area; there are complex rhythms in everybody.

We further mention the analysis by Başar et al. (1976) and Başar (1980) emphasizing the diversity of types of alpha rhythmicities (evoked and spontaneous, central or occipital). Accordingly, since the EEG electrodes are recording these compound potentials, the high-dimension D_2 between 5 and 8 only in the alpha frequency range should not be considered as very high: In the next chapter we will show that a much simpler system (smooth muscle) has a D_2 already in the order of 3.

In this survey a critical and complete review of alpha rhythmicities is not the aim. However, even this short review is enough to indicate the variety of the "alpha's." The large dimensionality of some alphas is not surprising. In my opinion (if I do not take into account the assumption of Babloyantz that dimensions above 8 cannot be considered as deterministic chaos), the data of Saermark et al. (1989) and Graf and Elbert (1989) are probably closer to reality for the entire EEG frequency range.

Experiments with proper physiological and psychophysiological design and step-by-step evaluation of D_2 together with simultaneous power spectral analysis might open new horizons for the understanding of alpha's. I also want to mention that 10-Hz rhythm activity should not be considered as a problem unique for higher vertebrates and one of pure cortical origin. (See, for example, Lopes da Silva et al. 1976; Bullock and Başar 1988).

Babloyantz' expression "strange-strange attractor" is probably not exaggerated since it can be shown (see Sect. 8 of this survey) that three simple smooth-muscle rhythmicities of isolated blood vessel already present a strange attractor. The EEG covers activity of a very high range of neural population. I think that it is a wonder to find a D_2 as low as 5 or 8 by scalp measurements.

6   Neuronal Oscillations, Evoked Rhythmicities, and Resonance in CNS

At the beginning of this survey I pointed out the importance of new investigations and results at the cellular level. I now come back to this point since the linkage of new results on the neuronal level might significantly extend the importance of results on chaotic dynamics of field potentials.

Freeman (1975) and Freeman and Skarda (1985) have shown that the EEG of the olfactory bulb and cortex in awake, motivated rabbits and cats has a characteristic temporal pattern consisting of bursts of 40- to 80-Hz oscillations, superimposed on a surface-negative baseline potential shift synchronized to each inspiration. These results were interpreted as follows:

> The neural activity which is induced by an order during a period of learning provides the specification for a neural template of strong connections between the neurons made active by that order. Subsequently when the animal is placed in the appropriate setting, the template may be "activated" in order to serve as a selective filter for search and detection of the expected odour (Freeman 1979).

> A key step in information processing depends on an orderly transition of cortical activity from a quasi-equilibrium to a limit-cycle (synchronized oscillation state) and back again. In this interpretation, the synchronized 40 Hz EEG activity, or the 40 Hz limit-cycle activity, serves as an operator on sensory input, to abstract and generalise aspects of the input into pre-established categories, thereby creating information for further central processing (Freeman 1983).

Recently Gray and Singer (1987, 1989; Gray et al. 1989) have reported that neurons in the cat visual cortex exhibit oscillatory responses in the range of 40–60 Hz. These oscillations occur in synchrony for cells located within a functional column and are tightly correlated with local oscillatory field potentials. This led Gray and Singer to the working hypothesis that the synchronization of oscillatory responses of spatially distributed, feature-selective cells might be a way to establish relations between features in different parts of the visual field. Later, Gray and Singer (1989) provided evidence that neurons in spatially separate columns synchronize their oscillatory responses. This synchronization occurs on the average with no phase difference and depends on the spatial separation and orientation preference of the cells.

The discovery made by Singer's group was later confirmed by Eckhorn et al. (1988, 1989a, b), who raised the important question whether coherent oscillations do reflect a mechanism of feature linking in the visual cortex. They also found stimulus-evoked resonances of 35–85 Hz throughout the visual cortex when the primary coding channels were activated by their specific stimuli. The results of the highly relevant experimental results by the groups of Singer and later Eckhorn and their interpretation was commented upon by Stryker (1989): "Is Grandmother an oscillation? Is it possible that the neurons in visual cortex activated by the same object in the world tend to discharge rhythmically and in unison? Such a one-note neural harmony could, in principle at least, provide the neurons at higher cortical levels with stronger inputs so that they associate the activities of lower-order neurons with one another." Stryker further notes, "Exploring the rhythms of the brain, revered by the pioneers of Electroencephalography but now mostly dismissed as irrelevant to neural information pioneering, may even come back into fashion."

The results of my group supported the results of Freeman and some of the interpretations. Furthermore, the possibility of generalization of 40-Hz evoked oscillations to other sensory modalities as well as a wider range of EEG frequencies has been discussed extensively in earlier publications (Başar et al. 1984, 1989 a, b; Başar and Stampfer 1985).

Although in the present survey I have given an example of analysis in the 8- to 13-Hz range of sustained oscillations (see Sect. 5), my coworkers and I tentatively concluded that activities of 1–4, 4–7, and 8–13 Hz serve as "operators" in the selective filtering of expected target stimuli. We suggested that Freeman's concept can be generalized to various sensory systems and to other EEG frequencies such as 2, 5–6, and 8–13 Hz. Furthermore, experiments in earlier studies have shown that a regular pattern of stimulation can induce a "preferred" phase angle, which appears to facilitate an optimal brain response to the sensory input.

We used the expression "operative states" for degrees of brain synchronization in defined frequency channels. In contrast to statements made by Freeman (1983), we suggested that EEG activity also contains response fragments to internal sensory and cognitive inputs and not only expectancy waves.

Is the use of the expression "EEG activity as a functional operator" in CNS a too general and strong assumption, or is the expression "operative states" for degrees of synchronization in various EEG frequencies a too courageous step? Since 1980 I have used the terms "alpha response" or "theta response" for evoked-potential components and published a general working hypothesis that all brain rhythms are linked to brain responsiveness (Başar 1980). Furthermore, two different oscillatory responses in human visual evoked potentials were described: (a) damped oscillatory responses of 5–6 Hz and (b) damped oscillatory responses of 10 Hz (Başar 1988c). Although the functional correlates of these two usually alternating different frequency responses are yet unclear, the results were interpreted that the brain had at least bistable responsiveness.

In a highly important recent review Llinás (1988) discussed the relevance of 6- and 10-Hz oscillations in the CNS. Llinás asks the following question: "How do the oscillatory properties of central neurons relate to the information-carrying properties of the brain as a whole?" Llinás (1988) described the thalamic neurons oscillating at two distinct rhythms: if the cell is depolarized, it may oscillate at 10 Hz; if the cell is hyperpolarized, it tends to oscillate at 6 Hz. Llinás concludes that oscillation and resonance allow single elements in the CNS to be woven into functional states capable of representing and embedding external reference frames into neural connectivity. In addition to these embedding properties, oscillation and resonance generate global states such as sleep-wakefulness rhythms and probably emotional and attentive states. This conclusion, which was reached by using results of the resonance concept at the neuronal level, has parallels to my conclusion that various cognitive tasks or sensory communications result in a specific combination of various "resonant modes," and that the ensemble of resonant modes could achieve an important function in the sensory communication of the brain (Başar 1980, 1988 c; see also Pfurtscheller et al. 1988).

What is the significance of results in studies of chaos in brain function? Does there exist a relation between findings at the neuronal level and the demonstration that EEG rhythms are not necessarily random stages?

The statement of Llinás that oscillation and resonance allow single elements in the CNS to be woven into a functional state supports the view of Freeman on 40-Hz activity and that of Başar on general resonance phenomena in theta and 10-Hz ranges. Although the designs of experiments were completely different in the studies mentioned, the coincidence of functional interpretation is highly striking.

Why do I return again and again to functional rhythmicities in the electrical activity in the brain? The legitimation of the expression "renaissance of EEG" (see Sect. 10) could not find unlimited support only with the help of analysis of D_2. The relevance of EEG in brain function can be demonstrated only by taking into consideration multidisciplinary approaches, including the results at the cellular level, in combined cognitive studies with the extension of the new scope of chaotic dynamics. Like Llinás (1988), I also think that the ability to project oscillatory rhythms and to generate synchronous firing in large populations of cells may belong to important properties of intrinsic electroresponsiveness.

7   High-Frequency Attractor in the Cerebellum and in Brain Stem

Although the spontaneous electrical activity of the brain which we call EEG has its most ample components in a frequency range between 1 and 50 Hz, there are a number of higher frequency components in the field potentials of the brain. Since the days of Adrian (1935) it has been well known that the cerebellum shows a high-frequency component between 180 and 300 Hz. According to the experiments described by Gönder and Başar (1977) and Başar (1980), who have shown resonance phenomena in a frequency range between 100 and 1000 Hz, Röschke and Başar (1989; and this volume) applied the Grassberger and Procaccia (1983) algorithm to this very high frequency activity. In this frequency region cerebellum and brain stem activity showed saturation in a number of experiments, and D_2 was in the range of $D_2 = 7$ for cat inferior colliculus and for cerebellum. The saturation was, however, reached only in about approximately 25% of the studied samples. Accordingly, Röschke and Başar (1989) concluded that in comparison to the SWS stage this is an unstable attractor. Since these data were obtained consistently in a number of experiments with four cats and during a long period of experimental time, it can be concluded that the brain also behaves in a deterministic manner during this frequency range, at least for some states of the brain. According to Röschke and Başar (1989) this behavior does not reflect noise and should be considered as a "very high frequency attractor."

What is the implication of this? Although studies undertaken in this frequency range are rare, we want to point out the brain stem responses that are used very frequently by neuroscientists. Besides the frequency range of 1–50 Hz, well known to the electroencephalographer, and also the new emerging field of the activity between 40 and 80 Hz (see Freeman 1975; Gray and Singer 1989; Eckhorn et al. 1989a, b; Galambos et al. 1981) there might be in future a new window for further investigations based on even higher frequency bands. The reader is also referred to the paper by Adey (1989).

8   Chaotic Activity of Smooth Muscle Shows Enormous Transitions: Consequences for Analysis of Brains Activity

I tentatively assume that besides the determination of D_2 one of the most important steps is the elucidation of the related function. Comparison of D_2 in functionally different states may lead to increased physiological understanding. For this purpose we leave, for a while, brain dynamics and consider more easily understandable problems in physiology. We turn our attention to vascular dynamics.

Smooth muscles show spontaneous contractions. At least three or four kinds of mechanical contractions of smooth muscle were classified by Golenhofen (1970) and by Bülbring and Shuba (1976). Later, in our research group the classification of smooth-muscle autonomous rhythmicity was characterized with spectral analysis (Başar-Eroglu et al. 1981). Smooth-muscle contractions are centered on a 1-min period (or 0.02 Hz), another rhythmicity of 0.06 Hz called basal organ rhythms, and faster rhythms (0.1 Hz).

Figure 6 shows spontaneous contractions of an isolated smooth-muscle preparation, namely the portal vein of guinea pig. In Fig. 6A there is a limit cycle type of contraction with a periodicity of about 60 s (minute-rhythm). The D_2 evaluated for this oscillation is $D_2 = 1.15$. The irregular motion shown in Fig. 6B represents a good saturation plateau (i.e., no significant noise) with a higher value of $D_2 = 3.3$, and the power spectrum of contractions contained three distinct peaks.

Now we ask the following question: The portal vein has three different contractile activities, and if these interact, we find a transition to a new chaotic stage with a D_2 of about 3. First of all, it is necessary to make the comment that in a blood vessel the physiological function of the smooth-muscle activity is correlated with vasomotion. Essentially the same activity was also obtained in isolated strips of smooth muscles in stomach, uterus, intestines, and all smooth-muscle organs. In most of the gastrointestinal organs, the activity is related to peristalsis, in other words, propulsion of nutritive elements. In comparison to brain function the function of smooth muscle is simple. However, we see in smooth muscle that there is also a transition from an almost limit cycle behavior ($D_2 = 1.15$) to a strange attractor behavior ($D_2 = 3.3$). If for relatively simple motions like vasomotion or peristalsis this type of transition is possible, where already a value of $D_2 = 3$ can be found, then it is not surprising that during the waking state of

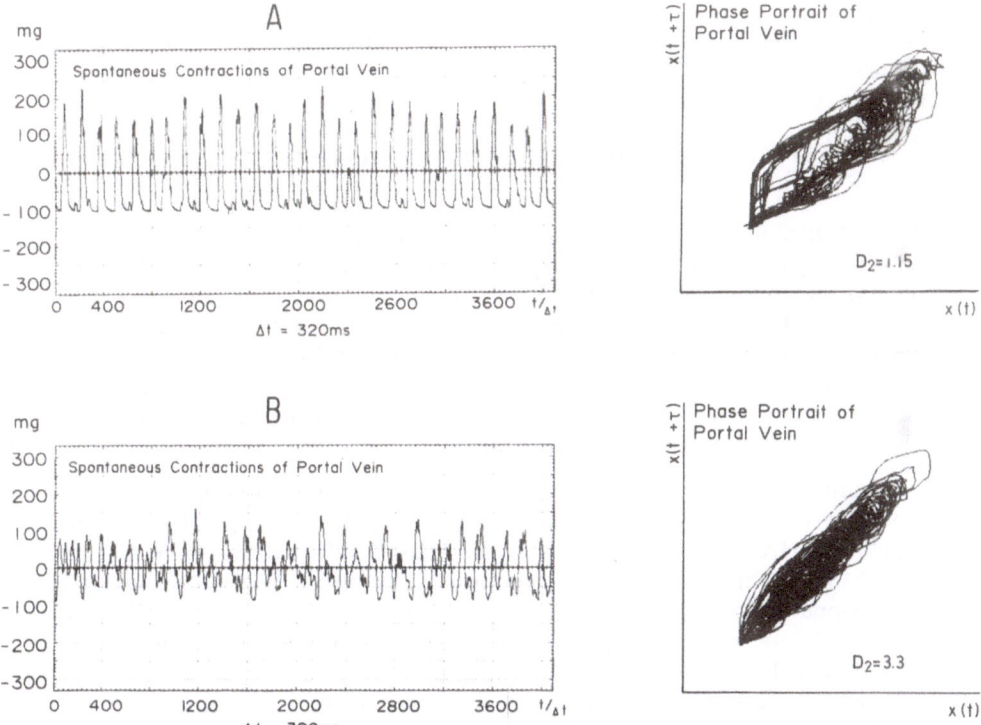

Fig. 6 A, B. Spontaneous mechanical contractions of the portal vein. *Left*, time series; *right*, phase portrait. (A) Example of the so-called "minute rhythm" activity. (B) A period in which the smooth muscle presents a mixed contractile activity. Compare changes in the dimensionalities in (A) and (B). (From Başar et al. 1986)

the brain values of 5–8 (the alpha activity) were obtained. The number of neurons involved in alpha activity and/or the number of inputs converging to this neural activity must be very high in comparison to the smooth-muscle activity.

If such a relatively simple system of two or three coupled oscillators – the smooth-muscle cells are at least mechanically coupled – may show such big transition changes (from 1 to 3), the transitions from 5 to 7.5 during alpha activity of the brain is also understandable (see Figs. 3, 4). Although here only the results of the guinea pig portal vein are shown, the same contractile activity is present in other types of organs from which the regulatory activity is dominated by a smooth-muscle effector (stomach, intestinus, uterus, lymph nodes, etc.; see Başar-Eroglu et al. 1981).

It is also noteworthy that Aladjalova (1964) described "infra slow" rhythmic oscillations with periods of 7–8 s and of 0.2–2 min in the brain. Are there mutual interactions between brain attractors and peripheral attractors in smooth-muscle organs as well as the vasculature? This question is difficult to answer. However, the view of Skinner et al. (1989) and Keidel et al. (1989a, b), looking for attractors in the integrative CNS cardiovascular system, should be taken into consideration by including hypothalamus, the autonomic nervous system and the peripheral sense organs.

9   Words of Caution: A Short Summary of Criticisms

As I mentioned in Sects. 2 and 3 of this survey, the new emerging branch of EEG and neuroscience confronts at the moment a number of difficulties that stem from (a) methodological concerns, (b) difficulties in designing proper experiments, and (c) overenthusiasm to estimate the D_2 for a number of brain structures and pathological cases without proper experimental design.

Since it is impossible as yet to provide a complete general agenda for precaution, I will try to give a brief summary of critical remarks which are additionally included in the various chapters of this volume:

1. According to the description of Babloyantz (this volume), Rapp et al. (1989), Albano et al. (1988), and Röschke and Başar (1989), length, stationarity, delay, and frequency resolution of signals must be adequately chosen.
2. As Mpitsos stated (1989), the estimation of the dimensionality by looking at the slope of curves and the criteria for convergence are still subjective. It would be desirable to develop universal subroutines that would check automatically the accuracy of convergence criteria.
3. Designs of experiments similar to the concepts of Freeman and Skarda (1985), Skinner et al. (1989), Mpitsos (1989), and Başar et al. (1990) are extremely useful to correlate behavior with chaotic activity.
4. As the analysis of Albano et al. (1987, 1988) or the analysis of smooth-muscle contractions described in Sect. 8 showed, it is useful to see what simpler systems do in order to evaluate brain activity.
5. Before comparisons, for example, of pathological states it is important to consider digitization noise, amplifier noise, analog, and digital filters used. Only data recorded with similar recording devices and measured with the same computer routines may be compared.
6. Comparison of the absolute dimensionality between two different brain states or recordings from various brain sites can be erroneous (Röschke and Başar 1989).
7. It is important to take into account the questions raised by Bullock (1989 b) about noise and amplitude of various rhythmicities. Evolution of the brain will be also an important aspect to study (see Sect. 4.9).
8. New algorithms are needed to extend that of Grassberger and Procaccia (1983). Kolmogorof entropy is still less used. While new and better algorithms are desirable, a possibility for computation of the dimension for shorter EEG segments must be sought. (See also new conceptual suggestions by Rapp et al. 1989).
9. Neuroscientists should avoid excessive enthusiasm about chaos and brain function and exaggerated speculative statements. This is in order to protect serious development in this field, which has already achieved important conceptual changes.

10 Important Messages

Some of the important messages of the chapters in this volume can be outlined as follows:

1. Renaissance of EEG. The brain may sometimes act as a noisy tissue, or a tissue with random activity. Sometimes it shows chaotic activity. This means that there are a number of brain structures and a number of brain states during which EEG is not noise but a strange attractor. This new evidence alone is an important breakthrough to future applications of EEG, especially in cognitive neuroscience. If the new trend can be used properly, this new window, in which some states of EEG will be considered as a "hot signal," the related research of EEG and magnetoencephalography, can develop to be even more relevant than it is now for the elucidation of brain function.
2. Transitions in brain waves (or bifurcations) can be considered as important indicators accompanying pathology in brain function and in cognitive information processing; this is reflected by adequate use of the correlation dimension.
3. The integrative function of the central nervous system seems to be a good approach for introducing the concepts of the attractors.
4. I highly recommend taking into consideration the agenda for research in chaotic dynamics that includes several crucial points neglected by several scientists (see Bullock, this volume).

How Far Can We Go? I am against chaotic statements in chaos science. An example of this is the article by Pool (1989). No doubt the author picked up interesting examples of scientists seriously working in this field. However, to me and to several colleagues, the interpretations are too early and too speculative. On the other hand, several statements were already possible by using tools of conventional physiology and neurophysiology without using the magic word "chaos." This type of article, verging on the sensational, might damage a smooth and healthy development in the field of nonlinear dynamics. In the same article it is stated that, "Healthy systems don't want homeostasis. They want chaos." Should we completely forget or deny Cannon's work? What should we then answer to such possible questions as: "Is then a normally beating heart unhealthy but a fibrillating is healthy?" or "Regular alpha waves are not healthy but chaotic-looking desynchronized EEG activity is healthy?" I am sure that avoiding premature statements will be healthy for the newborn child of "chaotic brain dynamics."

I think that new avenues can be opened in future research by application of several rules (some of them outlined above) and adequate design of experiments requiring basic physiological background and mathematical strength. Our way can be long, but fruitful, if we take into account necessary precautions and provide intelligent steps. The path can be very short if the nonlinear scientist's own enthusiasm forces them to publish every "seemingly bright idea."

11   Good News Is Bad News

The expression above probably reflects the most important wisdom of traders on Wall Street. If a stock's name is in every mouth, then it is time to be cautious before buying it. In recent years there has been only good news about chaotic dynamics, including highly promising comments about health, intelligence, and sudden deaths. It is impossible to refer to a large number of articles describing good news and unlimited opportunities to deal with biological systems with the microscope of chaotic attractors.

Although the new approaches have opened important conceptual new windows (some of them described briefly in the present volume), we must keep our eyes open not do damage this important new branch of neuroscience. Chaotic dynamics should not be developed to be a target of critics based on papers that contain no experimental information, unevaluated data, or data describing the behavior of only one subject or one case study. Methodological shortcomings that give rise to slight differences in experimental results and existing divergences among pioneering data can be corrected in the coming years. These discrepancies do not lead to contamination of the concepts. Better statistics and unified conventions (filters used, number of measuring points, sampling, amplifier noise, etc.) will certainly be established soon. However, superficial and exaggerated statements can create bad news for many years. Bullock (1984) noted that in neuroscience quiet revolutions are needed. I think, despite a number of difficulties, the new area treating "chaos in brain function" is one of the necessary quiet revolutions in neuroscience.

12   Summary

Dynamical analysis of brain electrical signals is a rapidly developing branch in neuroscience. Results from studies on physical systems demonstrated that very simple deterministic systems of differential equations can display highly disordered, turbulent behaviour. Since 1983 a number of neuroscientists and mathematicians have started to apply approaches from mathematical sciences to brain function, trying to extract information on physiology of behaviour from chaotic brain activity. This paper displays a survey of publications in this new development. The results and the experiments of chaotic dynamics with evaluation of correlation dimension both for human and animal physiological experiments were surveyed and discussed. Moreover, the author, who has been involved in this new branch of research for the last years tries to inject his own ideas about the recent steps by leaning on neurophysiological experiments. Accordingly, oscillatory phenomena of field potentials together with recent results on the cellular level are also strongly emphasized in this survey.

The discussion based on theoretical as well as experimental results has the following important message: The analysis of chaotic dynamics of the brain opened a new area, a type of "Renaissance" in brain research, in which the EEG analysis now gains an important insight. The strong emphasis of the paper is, however, to keep eyes open to correlate the findings of mathematical analysis with physiological function in an interwoven way and take precautions to avoid steps leading to several erroneous interpretations. Introductory remarks on concepts and methods are also included in order to orient the reader for the highly qualified pioneering work presented in the volume.

References

Abraham RH, Shaw CD (1983) Dynamics. The geometry of behaviour, vols 1–3. Aerial, Santa Cruz

Adey RW (1972) Organization of brain tissue: is the brain a noisy processor? Int J Neurosci 3:271–284

Adey RW (1989) Bioinstrumentation – cutting edge and limiting factor in the future of brain research. In: Başar E, Bullock TH (eds) Brain dynamics. Springer, Berlin Heidelberg New York, p 498

Adrian ED (1935) Discharge frequencies in cerebral and cerebellar cortex. J Physiol 83:32–33

Aladjalova NA (1964) Slow electrical processes in the brain. Progress in brain research, vol 7. Elsevier, Amsterdam

Albano AM, Mees AI, de Guzman GC, Rapp P (1987) Data requirements for reliable estimation of correlation dimensions. In: Degn H, Holden AV, Olsen LF (eds) Chaos in biological systems. Plenum, New York, pp 207–220

Albano AM, Muench J, Schwartz C, Mees AI, Rapp PE (1988) Singular-value decomposition and the Grassberger-Procaccia algorithm. Am Phys Soc 38(6):3017–3026

Babloyantz A (ed) (1986) Molecules, dynamics, and life. Wiley, New York

Babloyantz A (1988) Chaotic dynamics in the brain activity. In: Başar E (ed) Dynamics of sensory and cognitive processing of the brain. Springer, Berlin Heidelberg New York, pp 196–202 (Springer series in brain dynamics, vol 1)

Babloyantz A, Destexhe A (1986) Low dimensional chaos in an instance of epilepsy. Proc Natl Acad Sci USA 83:3513

Babloyantz A, Nicolis C, Salazar M (1985) Evidence of chaotic dynamics of brain activity during the sleep cycle. Phys Lett (A) 111:152–156

Başar E (1976) Biophysical and physiological systems analysis. Addison-Wesley, Reading

Başar E (1980) EEG-brain dynamics. Relation between EEG and brain evoked potentials. Elsevier/North-Holland, Amsterdam

Başar E (1983a) Synergetics of neuronal populations. A survey on experiments. In: Başar E, Flohr H, Haken H, Mandell AJ (eds) Synergetics of the brain. Springer, Berlin Heidelberg New York, pp 183–200

Başar E (1983b) Toward a physical approach to integrative physiology. I. Brain dynamics and physical causality. Am J Physiol 245:R510–R533

Başar E (1988a) Epilogue: brain waves, chaos, learning, and memory. In: Başar E (ed) Dynamics of sensory and cognitive processing by the brain. Springer, Berlin Heidelberg New York, pp 395–406 (Springer series in brain dynamics, vol 1)

Başar E (ed) (1988b) Dynamics of sensory and cognitive processing by the brain. Springer, Berlin Heidelberg New York (Springer series in brain dynamics, vol 1)

Başar E (1988c) EEG – dynamics and evoked potentials in sensory and cognitive processing by the brain. In: Başar E (ed) Dynamics of sensory and cognitive processing by the brain. Springer, Berlin Heidelberg New York, pp 30–55 (Springer series in brain dynamics, vol 1)

Başar E, Bullock TH (eds) (1989) Brain dynamics. Springer, Berlin Heidelberg New York (Springer series in brain dynamics, vol 2)

Başar E, Stampfer HG (1985) Important associations among EEG-dynamics, event-related potentials, short-term memory and learning. Int J Neurosci 26:161–180

Başar E, Gönder A, Ungan P (1976) Important relation between EEG and brain evoked potentials. II. A system analysis of electrical signals from the human brain. Biol Cybern 25:41–48

Başar E, Başar-Eroglu C, Röschke J, Schütt A (1984) A new approach to endogenous event-related potentials in man: relation between EEG and P300 wave. Int J Neurosci 24:1–21

Başar E, Başar-Eroglu C, Röschke J (1986) Dimensionality of smooth muscle contractions. Conference on perspectives in biological dynamics and theoretical medicine. The New York Academy of Sciences, abstracts book, p 4

Başar E, Başar-Eroglu C, Röschke J (1988) Do coherent patterns of the strange attractor EEG reflect deterministic sensory-cognitive states of the brain. In: Markus M, Müller Sc, Nicolis G (eds) From chemical to biological organization. Springer, Berlin Heidelberg New York, pp 297–306

Başar E, Başar-Eroglu C, Röschke J, Schütt A (1989a) The EEG is a quasi-deterministic signal anticipating sensory-cognitive tasks. In: Başar E, Bullock TH (eds) Brain dynamics. Springer, Berlin Heidelberg New York, pp 43–71 (Springer series in brain dynamics, vol 2)

Başar E, Başar-Eroglu C, Röschke J, Schult J (1989b) Chaos- and alpha-preparation in brain function. In: Cotteril R (ed) Models of brain function. Cambridge University Press pp 365–395

Başar E, Başar-Eroglu C, Röschke J, Schult J (1990) Strange attractor EEG as sign of cognitive function. In: John ER, Harmony T, Prichep L, Valdes-Sosa M, Valdes-Sosa P (eds) Machinery of the mind. Birkhäuser, Boston (in press)

Başar-Eroglu C, Demir N, Tümer N, Tümer A, Başar E, Weiss C (1981) Overall myogenic coordination in visceral organs and lymph nodes. In: Başar E, Weiss C (eds) Vasculature and circulation. Elsevier/North-Holland, Amsterdam, pp 227–253

Birkhoff GD (1932) Sur quelques courbes fermées remarquables. Bull Soc Math Fr 60:11–26

Bülbring E, Shuba MF (1976) Physiology of smooth muscle. Raven, New York

Bullock TH (1976) Redundancy and noise in the nervous system: does the model based on unreliable neurons sell nature short. In: Reuben JP, Purpura DP, Bennett MVL, Kandel ER (eds) Electrobiology of nerve, synapse and muscle. Raven, New York

Bullock TH (1984) Comparative neuroscience holds promise for quiet revolutions. Science 225:473–478

Bullock TH (1988) Compound potentials of the brain, ongoing and evoked: perspectives from comparative neurology. In: Başar E (ed) Dynamics of sensory and cognitive processing by the brain. Springer, Berlin Heidelberg New York, pp 3–18 (Springer series in brain dynamics, vol 1)

Bullock TH (1989a) The micro-EEG represents varied degrees of cooperativity among wide-band generators. In: Başar E, Bullock TH (eds) Brain dynamics. Springer, Berlin Heidelberg New York, pp 5–12 (Springer series in brain dynamics, vol 2)

Bullock TH (1989b) Epilogue: signs of dynamic processes in organized neural tissue: extracting order from chaotic data. In: Başar E, Bullock TH (eds) Brain dynamics. Springer, Berlin Heidelberg New York, pp 539–547 (Springer series in brain dynamics, vol 2)

Bullock TH, Başar E (1988) Comparison of ongoing compound field potentials in the brains of invertebrates and vertebrates. Brain Res Rev 13:57–75

Dvorak I, Siska J (1986) On some problems encountered in the estimation of the correlation dimension of the EEG. Phys Lett A 118:63–66

Eckhorn R, Bauer R, Jordan W, Brosch M, Kruse W, Munk M, Reitboeck HJ (1988) Coherent oscillations: a mechanism of feature linking in the visual cortex? Biol Cybern 60:121–130

Eckhorn R, Reitboeck HJ, Arndt M, Dicke P (1989a) Feature linking via stimulus – evoked oscillations: experimental results from cat visual cortex and functional implications from a network model. Conference on Neural Networks, Washington, abstract volume

Eckhorn R, Bauer R, Reitboeck HJ (1989b) Discontinuities in visual cortex and possible functional implications: relating cortical structure and function with multielectrode/correlation techniques. In: Başar E, Bullock TH (eds) Brain dynamics. Springer, Berlin Heidelberg New York, pp 267–278 (Springer series in brain dynamics, vol 2)

Freeman WJ (ed) (1975) Mass action in the nervous system. Academic, New York

Freeman WJ (1979) Nonlinear gain mediating cortical stimulus response relations. Biol Cybern 33:237–247

Freeman WJ (1983) Dynamics of image formation by nerve cell assemblies. In: Başar E, Flohr H, Haken H, Mandell AJ (eds) Synergetics of the brain. Springer, Berlin Heidelberg New York, pp 102–121

Freeman WJ (1989) Analysis of strange attractors in EEGs with kinesthetic experience and 4-D computer graphics. In: Başar E, Bullock TH (eds) Brain dynamics. Springer, Berlin Heidelberg New York, pp 512–520 (Springer series in brain dynamics, vol 2)

Freeman W, Skarda CA (1985) Spatial EEG-patterns, non-linear dynamics and perception: the neo-Sherringtonian view. Brain Res Rev 10:147–175

Galambos R, Makeig S, Talmachoff PJ (1981) A 40-Hz auditory potential recorded from the human scalp. Proc Natl Acad Sci USA 78:2643–2647

Gebber GL (1980) Central oscillators responsible for sympathetic nerve discharge. Am J Physiol 239:H143–H155

Gebber GL, Barman SM (1980) Basis for 2–6 cycle/s rhythm in sympathetic nerve discharge. Am J Physiol 239:R48–R56

Gebber GL, Barman SM (1981) Sympathetic-related activity of brain stem neurons in baroreceptor-denervated cats. Am J Physiol 240:R348–R355

Gebber GL, McCall RB (1976) Identification and discharge patterns of spinal sympathetic interneurons. Am J Physiol 231:722–733

Gleick J (1987) Chaos: making a new science. Penguin, New York

Golenhofen K (1970) Slow rhythms in smooth muscle (minute rhythm). In: Bülbring E, Shuba MF (eds) Smooth muscle. Arnold, London

Gönder A, Başar E (1978) Evoked frequency stabilization in the electric activity of the cat brain. Biol Cybern 31:193–204

Graf KE, Elbert T (1989) Dimensional analysis of the waking EEG. In: Başar E, Bullock TH (eds) Brain dynamics. Springer, Berlin Heidelberg New York, pp 174–191 (Springer series in brain dynamics, vol 2)

Grassberger P, Procaccia I (1983) Measuring the strangeness of strange attractors. Physica (D)9:183–208

Gray CM, Singer W (1987) Stimulus specific neuronal oscillations in the cat visual cortex: a cortical function unit. Soc Neurosci 404:3

Gray CM, Singer W (1989) Stimulus-specific neuronal oscillations in orientation columns of cat visual cortex. Proc Natl Acad Sci USA 86:1698–1702

Gray CM, König P, Engel AK, Singer W (1989) Oscillatory response in cat visual cortex exhibit inter-columnar synchronization which reflects global stimulus properties. Nature 338:334–337

Haken H (1976) Synergetics. An introduction. Springer, Berlin Heidelberg New York

Haken H (ed) (1983) Advanced synergetics. Springer, Berlin Heidelberg New York

Kaiser F (1988) Theory of non-linear excitations. In: Fröhlich H (ed) Biological coherence and response to external stimuli. Springer, Berlin Heidelberg New York

Keidel M, Keidel WD, Tirsch WS, Pöppl SJ (1987) Studying temporal order in human CNS by means of "running" frequency and coherence analysis. In: Rensing L, an der Heiden U, Mackey MC (eds) Temporal disorder in human oscillatory systems. Springer, Berlin Heidelberg New York (Springer series in synergetics, vol 36)

Keidel M, Keidel WD, Tirsch WS, Pöppl SJ (1989) Temporal pattern with circa 1 minute cycles in the running coherence function between EEG and VMG in man. (In press)

Layne SP, Mayer-Kress G, Holzfuss J (1986) Problems associated with dimensional analysis of electroencephalogram data. In: Mayer-Kress G (ed) Dimensions and entropies in chaotic systems. Springer, Berlin Heidelberg New York, p 246

Llinás RR (1988) The intrinsic electrophysiological properties of mammalian neurons: insights into central nervous system function. Science 242:1654–1664

Lopes da Silva FH, Van Rotterdam A, Barts P, Van Heusden E, Barr W (1976) Models of neuronal populations: the basic mechanisms of rhythmicity. Prog Brain Res 45:281–308

Lopes da Silva FH, Kamphuis W, van Neerven JMAM, Pijn JPM (1990) Cellular and network mechanisms in the kindling model of epilepsy: the role of GABAergic inhibition and the emerge of strange attractors. In: John ER, Harmony T, Prichep L, Valdes-Sosa M, Valdes-Sosa P (eds) Machinery of the mind. Birkhäuser, Boston

Mees AI, Rapp PE, Jennings LS (1987) Singular-value decomposition and embedding dimension. Physiol Rev A36:340–346

Mpitsos GJ (1989) Chaos in brain function and the problem of nonstationarity: a commentary. In: Başar E, Bullock TH (eds) Brain dynamics. Springer, Berlin Heidelberg New York, pp 521–535 (Springer series in brain dynamics, vol 2)

Pfurtscheller G, Steffan J, Maresch H (1988) ERD mapping and functional topography: temporal and spatial aspects. In: Pfurtscheller G, Lopes da Silva FH (eds) Functional brain imaging. Huber, Toronto, pp 117–130

Poincaré H (1892) Les méthodes nouvelles de la mécanique céleste. Gauthier-Villars, Paris

Pool R (1989) Is it healthy to be chaotic? Science 243:604–605

Rapp PE, Zimmermann ID, Albano AM, Deguzman GC, Greenbaum NN (1985a) Dynamics of spontaneous neural activity in the simian motor cortex: the dimension of chaotic neurons. Phys Lett (A) 110:335–338

Rapp PE, Zimmermann ID, Albano AM, de Guzman GC, Greenbaum NN, Bashore TR (1985b) Experimental studies of chaotic neural behaviour: cellular activity and electroencephalographic signals. In: Othmer HG (ed) Nonlinear oscillations in biology and chemistry. Springer, Berlin Heidelberg New York, pp 175–205

Rapp PE, Albano AM, Guzman GC, Greenbaum NN, Bashore TR (1986) In: Othmer HG (ed) Nonlinear oscillations in biology and chemistry. Springer, Berlin Heidelberg New York, p 175 (Lecture notes in biomathematics, vol 66)

Rapp PE, Bashore TR, Martinerie JM, Albano AM, Mees AI (1989) Dynamics of brain electrical activity. Brain Topography (in press)

Röschke J (1986) Eine Analyse der nichtlinearen EEG-Dynamik. Doctoral dissertation, University of Göttingen

Röschke J, Başar E (1985) Is EEG a simple noise or a "strange attractor"? Pflügers Arch 405:R45

Röschke J, Başar E (1988) The EEG is not a simple noise: strange attractors in intracranial structures. In: Başar E (ed) Dynamics of sensory and cognitive processing by the brain. Springer, Berlin Heidelberg New York, pp 203–216 (Springer series in brain dynamics, vol 1)

Röschke J, Başar E (1989) Correlation dimensions in various parts of cat and human brain in different states. In: Başar E, Bullock TH (eds) Brain dynamics. Springer, Berlin Heidelberg New York, pp 131–148 (Springer series in brain dynamics, vol 2)

Rosen R (1969) Hierarchical organization in automata theoretic models of the central nervous system: In: Leibovic KN (ed) Information processing in the nervous system. Springer, Berlin Heidelberg New York

Rössler OE, Hudson JL (1989) Self-similarity in hyperchaotic data. In: Başar E, Bullock TH (eds) Brain dynamics. Springer, Berlin Heidelberg New York, pp 113–121 (Springer series in brain dynamics, vol 2)

Saermark K, Lebech J, Bak CK, Sabers A (1989) Magnetoencephalography and attractor dimension: normal subjects and epileptic patients. In: Başar E, Bullock TH (eds) Brain dynamics. Springer, Berlin Heidelberg New York, pp 149–157 (Springer series in brain dynamics, vol 2)

Schuster HG (1988) Deterministic chaos. VCH, Weinheim

Skarda CA, Freeman WJ (1987) How brains make chaos in order to make sense of the world. Behav Brain Sci 10:161–195

Skinner JE, Reed JC (1981) Blockade of a frontocortical-brainstem pathway prevents ventricular fibrillation of the ischemic heart in pigs. Am J Physiol 240:H1156–H1163

Skinner JE, Martin JL, Landisman CE, Mommer MM, Fulton K, Mitra M, Burton WD, Saltzberg B (1989) Chaotic attractors in a model of neocortex: dimensionalitites of olfactory bulb surface potentials are spatially uniform and event related. In: Başar E, Bullock TH (eds) Brain dynamics. Springer, Berlin Heidelberg New York, pp 158–173 (Springer series in brain dynamics, vol 2)

Storm Van Leeuwen W (1977) The alpha rhythm. In: Cobb WA, van Duijn H (eds) Contemporary clinical neurophysiology. Electroencephalogr Clin Neurophysiol Suppl 34:1–7

Stryker MP (1989) Is grandmother an oscillation? Nature 338:297–298

Van Erp MG (1988) On epilepsy: investigations on the level of the nerve membrane and of the brain. Proefschrift Rijksuniversiteit, Leiden

Walter WG (1964) The convergence and interaction of visual, auditory, and tactile responses in human nonspecific cortex. Ann NY Acad Sci 112:320–361

An Agenda for Research on Chaotic Dynamics

T. H. BULLOCK

1   Introduction

A physiologist innocent of mathematics rushing in where wise men fear to tread may well be foolish, but he may nevertheless contribute by pointing out where the emperor has no clothes. The purpose of this commentary is to indicate the controls and calibrations that a brain physiologist considers first-order experiments when a new measure is proposed as a descriptor of some aspect of brain function, or to generalize the prescription when any measure is proposed as a descriptor of any real system.

Attention is focused, of course, on those controls and calibrations that appear to be insufficiently used or not yet reported, but which are necessary simply for describing the system, apart from any speculations about functional significance or mechanism. As will soon be obvious, this critic is not only innocent of mathematics but in consequence has the most superficial familiarity with the literature of chaotic dynamics. He relies on the most recent publications, particularly those aimed at brain function, such as the chapters of this book, to present or to cite whatever controls and calibrations are appropriate, whether self-made or made in classical papers on other systems.

The field is too new, as applied to biological systems, to assume, without citation, that all the controls have been done, are accepted as adequate, and are well known.

2   Review of the Knowledge of Brain Tissue as a System of Generators

It seems appropriate to begin with a brief recapitulation of the present understanding of the real system to which methods of characterizing the dynamics are applied. The picture most germane to procedures sensitive to mixtures and interactions of ongoing fluctuations, whether rhythmic or not, is this. The tissue consists of enormous numbers (ca. 5×10^5 neurons/mm³ in cortex) of microscopic generators which are *mainly subcellular* (each synapse, dendrite, axon collateral, terminal arbor, etc.), not simple dipoles, and not oriented in a predictable way relative to the gross axes of the soma, except sometimes on the average. Cells within a fraction of a millimeter are diverse, even opposite in response type to a given stimulus. Only in some places are they grossly aligned or geometrically ordered. Tissue separated by a few millimeters or centimeters is generally quite different in fine structure and in function. Each generator has a wide-band power spectrum (<0.1 to >1000 Hz), roughly similar to that of the compound field;

each is largely independent but synchronized in varying degrees, interacting with some of the others in mainly nonlinear ways, probably in a number of distinct ways, and these ways change dynamically in fractions of a second, as well as in minutes, hours, and longer. Typical neurons receive many thousands of pulsatile inputs per second from thousands of converging inputs, from dozens of types of sources. The fluctuations in interaction, in synchrony, and in phase relations reflect a cooperativity with large numbers of determinants (inputs and state variables) and many response (output) variables. Tissue volumes of a fraction of a cubic millimeter are commonly quite coherent with neighboring masses less than 1 mm away but have low coherence across a few millimeters, except for a few special driving rhythms.

I leave open such questions as whether true oscillations are a general or a special case, the possibility of resonance, whether evoked potentials arise from the same fraction of the population as the ongoing EEG or recruit largely new generators, and the like.

This picture differs from many models of the generation of EEG (usually implied rather than explicit), which intentionally simplify, or view the brain through the scalp, or omit words that should be expressly included, such as: "generators act *as though they were* resonating oscillators." The facts condensed into the foregoing picture sound like arguments against any "field" theory of brain operation, yet they really only place constraints upon the scale of the fields and the generality of the information carrying role of fields.

Some points of special relevance to dynamical analysis are the following:

1. Inhomogeneity of structure and function is marked between regions of the brain centimeters or millimeters apart, and it increases as the scale or volume of tissue considered decreases.

2. The elementary events of the EEG are not uniform synaptic potentials or events with one dimension versus time. They include synaptic events of diverse form in three-dimensional space in elaborate arbors, plus hyperpolarizations of hundreds of milliseconds, plus slower processes in dendrites and somata, plus plateau potentials, and different kinds of pacemaker potentials, spikes distributed in three-dimensional arbors including electrotonically decrementing spikes in axonal terminals. In addition, impedance changes of unknown variety, not to speak of influential interactions by peptide modulators, glia, and probably other components, contribute to a fluctuating state.

3. Any measure that depends on cellular and subcellular generators, coupling functions, and events reflects a much smoothed resultant when applied to the volumes seen even by extracellular microelectrodes.

4. Synchrony is not a unitary, well-understood parameter but a complex of phenomena, rarely distinguished from each other, let alone quantitatively evaluated. In fact synchrony is almost never verified but is invoked by eyeball judgement of raw records.

The concept of synchrony can have quite different meanings. In the context of electrocorticograms, having a broad-band spectrum within which all frequencies are simultaneously present and separately waxing, waning, and shifting phase (Bullock 1988), the concept does not require agreement in phase throughout the population but a fixed phase for each frequency. Quantification should express either the fraction of the contiguous population or the volume of neural tissue that shows a degree of agreement at stated frequencies. One cannot estimate synchronization in an EEG recording by inspection of single-channel time records or from the cross-correlation between two points. It could be found that two such points are highly correlated because of a common driver, yet the assemblage of tens of thousands of cells in a cubic millimeter at each locus might be very little synchronized, or it could be found that two loci are not correlated in the amplitude fluctuations of many frequencies yet the phases might be well synchronized over some band of frequencies.

The significant definition of synchrony for neural activity is any level of congruence among loci, above that of coincidence, in a volume of tissue, or in some fraction of the neighboring generators, within a defined frequency band. Coherence is a suitable, though imperfect, estimator (a) except for frequencies, usually very high and weak, dominated by pulsatile events with jitter and (b) providing it is computed for a contiguous volume rather than for two distant points. Coherence, like synchrony, can decline with distance quite unequally in different directions, e.g., radially versus tangentially in the cortex. The tangential distribution of this coherence for superficial laminae or integrated across the cortex, would be a useful measure of cortical synchrony for many purposes. The plot of coherence against distance between a number of closely spaced recording loci provides a measure for each frequency, is independent of the absolute amplitude in each channel, and of the particular phase; it does not lump all frequencies as cross-correlation does (Bullock and McClune 1989).

We have no means, even by modeling, to specify the degree of coincidence expected by chance in the theoretical case of no synchrony, that is, all generators being independent, so that we cannot say when a small degree of synchrony is above the expectation of the null hypothesis. This is because the known generators, let alone those still unknown, are not simple dipoles and are not oriented in a readily knowable fashion, nor is the volume conductor isotropic. Besides improvements in assessment, we need testable hypotheses for mechanisms of synchronization, of which there are no doubt more than one and probably largely nonlinear.

5. The terms and the concepts of "rhythm" and "oscillator" are widely used and familiar. There is no question but that they are heuristic, and if we would always say "treating the compound field potentials of the brain *as though they were* a mixture of rhythms arising in oscillators ...," one could not object. We have, however, little justification for proposing that the primary generators of EEG and EP are in fact oscillators, as a rule, not just exceptionally, meaning that their characteristic, synchronized activity is periodic in the 0.1- to 50-Hz range, in the general case. This is true even for the cases of relatively sharp power spectrum peaks, which are in any event special cases, since most micro-EEG recorded in various species and places in the brain lack sharp spectral peaks. This means that terms such as "oscillation" and "resonance" should for the present be accompanied by qualifying terms such as apparent, ostensible, presumed, or as though.

6. These problems are parts of the broader area of relative ignorance, in spite of heroic work by many investigators over many years, namely the semi-microstructure of the ongoing and the evoked compound field potentials, in the tangential as well as the radial dimensions, on a millimeter scale or finer and a fraction of a second time resolution – as called for in Bullock (1989).

3   Requirements for a Descriptive Measure

1. Reproducibility. In order to be a candidate for characterizing a system of interest, any measure must be shown to be reproducible to within stated limits under stated conditions. For the time being, if the measure is too laborious, time consuming, or expensive to repeat many times, it cannot be such a candidate. Many of the reports of dimensionality of the EEG give inadequate data for statistical evaluation.

2. Sensitivity to Contamination. Even if a measure is reasonably reproducible under the conditions of one investigator, who, like everyone else, has no practical way of independently assessing the amount of contamination present, it is part of the first order of business to demonstrate quantitatively how tolerant or intolerant the measure is to known amounts of common contaminants injected under control.

If the curve of correlation dimension (D_2) against embedding dimension saturates to a clearly horizontal plateau, one is permitted to say that there is negligible random, indeterminate activity contaminating the deterministic chaotic activity. "Noise" is a commonly used but unfortunate jargonistic term for this, since its common sense and first dictionary definition is unwanted interference with signals. Yet such noise may be highly structured and may be highly determinate or even limit cycles and would not prevent a good plateau. The contamination that prevents saturation in D_2 estimation is not simply "white noise;" it may be filtered and not "white," and it may not be interference but a true signal. It should be designated by a term or phrase that underlines its random, indeterminate nature. In the brain intrinsic random activity does not necessarily mean interference and is sometimes demonstrably useful.

The statement that EEG is chaotic, meaning that it saturates to a certain D_2, is a strong statement and a remarkable discovery, since it says that there is negligible random activity. We could not heretofore have said any such thing and would have doubted any claim to this effect. As Mpitsos (this volume) points out, instead of assuming, as we have, that deterministic signals ride on a background of random variation, it is now possible to show, in given cases, that the latter is insignificantly small and the observed fluctuation is essentially deterministic. But, how strong a statement is it; how sensitive is the method? Since we have no independent method of estimating the proportion of random activity, the practical value of D_2 claims depends on experimental tests with known proportions injected. This in turn depends on objective assessment of saturation – which appears to be rare.

Röschke (1986) implies that extrinsic "white noise" has a small effect on the estimate of correlation dimension if the signal-to-noise ratio is above 10. It is not clear how this dependence changes with D_2, and whether this is only a matter of scaling. It would also be important to learn whether nonwhite random activity, for example, a band of high frequency or of low frequency, exerts the same effect, according to its proportion of the total power.

3. Sensitivity to Nonstationarity. Besides contamination and reproducibility, a measure that integrates over time must be evaluated in terms of the degree to which it is influenced by trends or segments of the sample different from other segments by relevant criteria. Stationarity cannot be dismissed simply on the ground that it is a subjective judgement depending on the purposes of the investigator. It is at least incumbent on the investigator to state that the samples were examined for stationarity on criteria relevant to those purposes, and what the criteria were. Mpitsos (this volume) underlines the importance of this factor and thinks there is hope that with proper mathematical attention we can learn to handle time series that are clearly nonstationary, instead of simply rejecting such data, often of special interest. Havstad and Ehlers (1989) have dealt with nonstationary systems and claim to get statistically significant results with very small data sets. A reasonable agenda for learning the value of the new measure would include experiments in which the whole sample of data to be analyzed is composed of mixtures of segments of a known degree of difference, both in D_2 and in the criteria for stationarity.

4. Sensitivity to Single Components. It could be of some importance to know how large must be the amplitude of some component in order to exert an effect on D_2. For example, when does an incommensurable periodicity newly appearing, and independent of the rest of the components, reach a threshold amplitude and add its contribution to D_2 – in the presence of an inevitable but tolerable (Sect. 3.2, above) random component and of periodicities of large amplitude? How much effect does a single strong rhythm have, presumably suppressing D_2 by pushing some of the weaker components of the signals below threshold as a proportion of the total power? For example, does the low-amplitude ca. 40-Hz component of the EEG contribute less to the dimensionality in the case of a sample record that has high-amplitude 11-Hz (alpha) or 2-Hz (delta) than in the case of a record without such a power peak? This, again, is easily studied by experiments with controlled time series. It overlaps with point 2, above, in that the injected component might be a large amplitude limit cycle or independent chaotic series considered as a contaminant, e.g., an electrocardiogram or electrotonically spread activity from another part of the brain, buried in the EEG.

This is a logical place to insert the suggestion made to me by J. Theiler (personal communication) for a control experiment he would advocate as a standard procedure. (a) From the input time series, compute the Fourier spectrum. (b) From the Fourier spectrum create several new time series with the same spectrum. (One way is to add random phases to the spectrum and then invert the Fourier transform.) (c) Compute the dimension estimate for the new series and compare it with that for the original. If they are the same, to within the precision defined by the standard deviation of the ensemble of new series, there is no reason to believe that the original time series contains anything extraordinary or "nonlinear."

5. Sensitivity to the Judgemental Constants. Başar, Babloyantz, Rapp, and others have emphasized the importance of correct choice of digitization rate, number of data points, delay value, and filters. The very fact that these are to some extent judgemental sets the requirement that readers be informed how much effect a different choice would have, when applying the standard algorithms (Theiler 1986, 1990; Albano et al. 1987).

It should not be overlooked that some of them can be made substantially less subjective, according to findings of Mees et al. (1987) and Rapp et al. (1988). Clearly the methods are not yet acceptably standardized and routine; there are still major differences among the users of these methods in the criteria for selecting values, in the practical compromises and rules of thumb, as well as in their statements of the upper limit of reliable D_2 values. General agreement on such matters will come slowly and unevenly, one imagines. There is hope for significantly modified new algorithms to replace the Grassberger and Procaccia-based procedures now in general use (Rapp et al. 1990).

4   Evaluation of a New Measure

1. Calibration Against Familiar Systems. Beyond these basic questions of robustness, one looks forward to help in evaluating the numbers. When is D_2 high? Compared to what? One wonders, for example about the dimensionality of a microphone record of a voice that goes from singing to babbling to reading Shakespeare; or of several voices that go from unison recital to two and then three or more speaking independently at once? In my sampling of the literature and conversations with experts it is difficult to find analyses of more or less familiar systems of different D_2. Turbulence in a certain vessel of water, I am told (by J. Theiler; see Brandstater and Swinney 1987), caused by counterrotating cylinders, appears to have a correlation dimension of 3.3–5.3, depending on the velocity of stirring. Other systems for which we may have some intuitive feel are, however, almost never examined. I cannot find the dimensionality of the predicted tides of the ocean, let alone the observed sequence of sea levels, with the added sources of fluctuation. Cardiac arrhythmias would be a useful, well-known database to examine. A. T. Winfree points out (personal communication) the interesting possibilities of contrasting the phenomena as seen by the electrocardiogram, by surface electrodes, and by intracellular micropipettes. Kaplan (1989) tested certain kinds of experimentally induced fibrillation and models of fibrillation and found that they did not plateau so that if there is a chaotic attractor there is also a substantial random component.

Some experts in these methods profess not to trust measurements of actual physical systems, especially those with apparent fractal dimensionality above some very low value, because of pitfalls in the application of the algorithms.

Models simulating EEG with different mixtures of rhythms and nonrhythms, and different numbers of nonlinear interactions, frequency modulations, and amplitude modulations would help greatly to educate our intuitions about interpreting D_2 values. What specifications must a simulated system have to achieve a D_2 of 3.2, 3.8, 4.5, 5.5, etc.? Modeling can also give us some experience with the problem of time series reflecting large numbers of parallel generators and processes. Başar (this volume) compares smooth-muscle recordings and brain recordings. We have, however, no reliable intuition or experience that says "if 10^5 muscle cells give $D_2 = 1.15$ or 3.3, then 10^7 or 8 or 9 neurons should give $D_2 = 5$ or 8;" but this is subject to experiment by simulation.

2. Comparison Among Neural Systems and Volumes of Tissue Sampled. To this observer the currently reported correlation dimensions of 5–10 for the mammalian cortex, insofar as this implies about that many variables, must testify to a great smoothing, in space or time or both, in the light of the astronomical numbers of generators, the hundreds or thousands of cell types (Bullock 1980), and dozens of qualitatively different integrative variables (Bullock 1984) determining their activity.

Looking ahead, it is to be hoped that the power of these analyses will be extended to records of the same piece of brain with larger and smaller electrodes that see larger and smaller populations of cells. It would contribute greatly to an appreciation of D_2 values if we had them for simultaneous records from several scalp electrodes, placed closer together than usual, from several epidural electrodes, sealed into the skull in the same animal and region, permitting the comparison of pairs with high and pairs with low coherence, and from two or more intracortical extracellular microelectrodes. One channel of intracellular recording would add much more than simply frosting to the cake.

If, as seems possible, D_2 is not systematically lower as the volume of tissue sampled shrinks, one will have to conclude that the measure from the scalp or from any electrode is not in fact giving us any approximation of the number of incommensurable rhythms and interactions but is simply some resultant of the enormous smoothing, spatial averaging, dilution of signals by other signals, and other unknown consequences of the large number of parallel elements.

The preliminary reports available so far (see references in Başar 1988; and the chapters in the present volume by Röschke and Başar; Babloyantz; Rössler; Skinner et al.; Mpitsos; plus papers by Rapp et al. 1985; Hayashi and Ishizuka 1987; Labos 1987) encourage the search for this descriptor in various parts of the nervous system, including intracellular records not only of invertebrates (my old love, the follower cells of the cardiac ganglion of lobsters should be of special interest) but also of cortical, thalamic, limbic, reticular, cerebellar, and spinal neurons and glia, in resting, attentive, and striving states.

Since single-channel records provide such a meager sample of the spatiotemporal distribution of activity in the brain, it is to be hoped that the methods will be applied to recordings with 2, 3, 5, or 50 channels of simultaneous recording from neighboring parts of the brain, even though this increases the already formidable methodological problems. We cannot settle for the equivalent dipole so useful in cardiac physiology, since the imposed simplification would be too great, apart from special cases. In the general case we do not have a dominant rhythm like the alpha or theta, which are special cases, but a broad spectrum. We know that coherence falls with distance in the millimeter range, to a widely varying degree (Bullock et al. 1984; Bullock and McClune 1989; Bullock et al. 1990), but is appreciable even to centimeters in some frequency bands in the cases of driven rhythms. Cooperative among neurons is rife – and ripe for quantitative analysis. Invertebrate ganglia offer prime targets for methods sensitive to nonlinear interactions among a few dozens of cells. Not only spike intervals, but also slow, intracellular as well as extracellular field potentials need to be examined in invertebrates, since there is ample evidence that subthreshold, slow events contribute to interactions among neurons (Graubard et al. 1983).

3. Relation of D_2 to Complexity. It is clear that correlation dimension is not a monotonic measure of complexity of causes, in a commonsense meaning. For our purposes, *complexity can be defined* rigorously or at least measured by the number of bits of information necessary to characterize the system consistently (these words are necessary because a random series would require many bits to describe in its unique realization) or by the number of component parts and relations between them. A system with four independent, incommensurable rhythms has $D_2 = 4.0$, but one with the same plus some nonlinear interactions between two of them, hence a more complex system, may have $D_2 = 3.9$. Merely by changing the value of some exponent it might go to 3.3, without changing the complexity – as distinct from the elaboration of the resulting pattern. The sequence of whole numbers of course scales complexity as more independent rhythms are added. Perhaps it is true that between whole numbers, fractional dimensions sometimes scale complexity (D_2 of $3.9 > 3.8 > 3.7 > 3.6$ in complexity). It would be helpful to have experimental demonstrations with simulated data sets of known complexity together with any qualifications, e.g., how much can the change of a single constant or coefficient change the range of fractal dimensions? Does it follow that $3.9 < 4.9 < 5.9$ in the scale of complexity, as defined? Is then $3.9 < 4.1$ in complexity – so that we have a scale of complexity except that the whole numbers cannot be used in the scale? Authoritative assertions from theorists should help in the use of this measure. Unfortunately, there are now many self-proclaimed authorities who give confident statements at variance with each other. This particular question, more important to the users and potential users of the tools than to the designers, may be an example of a field of mathematics in which demonstration with known test data will be more convincing than argument from principles. What then does a higher D_2 mean? What does *more* chaotic mean?

One sees quotations about chaos analysis having shown that the brain is "more ordered" in one state than another. Is dimensionality a scale of orderedness? I take it this is intended to apply only to the fractals, not the whole numbers; a system of $D_2 = 4.0$, having four independent, incommensurable rhythms is not more *ordered* than one with 3.0, surely. If it applies to the fractions, is then a D_2 of 3.8 less ordered than one of 3.2? Is 3.5 more ordered than 4.5? Is 3.1 more ordered than 4.1? Do we really understand the alternative kinds of nonlinear coupling that can give the same D_2 well enough to assert that they are equal in orderedness? Or is this word even less easily defined than complexity? Whereas some uses of this term can be excused as being semipopular, it does appear in scientific literature, without definition as far as I can find, and clearly different in meaning among authors. It would seem wise either to avoid the term of to define it rigorously.

It is not necessary to wait for such questions to be settled to employ the methods of computing dimensionality as potentially valuable descriptors. The brain is unlike many other systems where we have little basis for anticipating the number of interacting variables or their linearity, since we already know many basic nonlinearities in the neural mechanisms of integration among cells and the very large numbers of quasi-independent units. We do not expect any magic new method to reveal the way the brain works or unlock a single code. Instead we need more descriptors with which to establish empirical correlations among descriptors and to measure the effect of perturbations, to help in system analysis. The wealth of questions explicitly formulated above, as well as those waiting to be formulated, give promise of an exciting future, once the methods of assessment are calibrated against simulated systems and controlled against unwanted variables.

We cannot yet say that all brains at all times have strange attractors, and non-random order, of integral or fractal dimensions. Albano et al. (1987) and an unreported number of experiments in various laboratories obtain the result quite often that a given data set representing a time series recorded from the brain does not resolve into a stated number of dimensions and cannot be characterized in these terms. But chaotic attractors are at least widespread, and their dimensionality may yet be found to correlate in some degree with higher species, levels of the CNS, and levels of cognitive performance. I, for one, anticipate an exciting period in the near future, while empirical experience accumulates and interpretations are disputed and refined. As one of a set of new descriptors, quite nonintuitive and rooted in the dynamic processes ongoing in the system, it will surely stimulate new experiments, provide surprises, lend support to some old prejudices and undermine others.

5 Summary

This essay is a call for controls and calibrations in reporting the results of dynamical analysis as descriptors of real systems. Too much of the literature on correlation dimensions, D_2, computed for brain activity lacks basic statistics and evaluation of the influence of various factors known to affect such algorithms.

Before detailing these needs, a short review is given of the present understanding of brain tissue as a system of astronomical numbers of generators, mainly subcellular. Points particularly relevant here are these. (a) Inhomogeneity of structure and function is marked and increases as the volume under consideration decreases. (b) The elementary events of the EEG are not uniform synaptic potentials but include events of diverse form in three-dimensional space in elaborate arbors. Fluctuations of state also arise from peptide modulators, glia, and other components. (c) Measures derived from electrodes that sample a cubic millimeter or more reflect a much smoothed resultant of large numbers of weakly coupled or quasi-independent sources. (d) Synchrony of several kinds varies widely in degree. (e) Rhythms and oscillations are not the general case but special cases, in certain places and conditions; frequencies isolated by filters or Fourier analysis do not necessarily signify oscillators. EEG power may be a mix of more or less regular and more or less synchronized cellular oscillations and discrete events, such as synaptic potentials.

Research on chaotic dynamics of brain activity as a potentially characteristic descriptor needs to satisfy the usual requirements of a measurement by demonstrating the statistical limits of repeatability, the sensitivity to contamination, to nonstationarity and to large amplitude single components. Those judgemental constants in the algorithm should be justified by objective evidence.

In addition, this new form of measurement needs calibration against more or less familiar systems and against simulated data sets whose composition is manipulated in specified ways. Measurements of actual neural tissue should systematically compare recordings from smaller and larger electrodes "looking at" smaller and larger populations of neurons, from lower and higher levels of the CNS, earlier and later ontogenetic stages, and at primitive and advanced species.

References

Albano AM, Mees AI, de Guzman GC, Rapp PE (1987) Data requirements for reliable estimation of correlation dimensions. In: Degn H, Holden AV, Olsen LF (eds) Chaos in biological systems. Plenum, New York, pp 207–220

Başar E (ed) Dynamics of sensory and cognitive processing by the brain. Springer, Berlin Heidelberg New York

Brandstater A, Swinney HL (1987) Strange attractors in weakly turbulent Couette-Taylor flow. Phys Rev A 35:2207–2220

Bullock TH (1980) Reassessment of neural connectivity and its specification. In: Pinsker HM, Willis WD Jr (eds) Information processing in the nervous system. Raven, New York, pp 199–220

Bullock TH (1984) A framework for considering basic levels of neural integration. In: Reinoso-Suarez F, Ajmone-Marsan C (eds) Cortical integration: basic, archicortical and association levels of integration. Raven, New York, pp 27–36

Bullock TH (1988) Compound potentials of the brain, ongoing and evoked: perspectives from comparative neurology. In: Başar E (ed) Dynamics of sensory and cognitive processing by the brain. Springer, Berlin Heidelberg New York, pp 3–18

Bullock TH (1989) The micro-EEG represents varied degrees of cooperativity among wideband generators: spatial and temporal microstructure of field potentials. In: Başar E, Bullock TH (eds) Brain dynamics. Springer, Berlin Heidelberg New York, pp 5–12

Bullock TH, McClune MC (1989) Lateral coherence of the electrocorticogram; a new measure of brain synchrony. Electroencephalogr Clin Neurophysiol (in press)

Bullock TH, Lange GD, McClune MC (1984) A measure of synchrony in the cortical EEG: the slow wave drowsy state is slightly more synchronized horizontally than the low voltage fast state. Soc Neurosci Abstr 10:1143

Bullock TH, Buzsáki G, McClune MC (1990) Coherence of compound field potentials reveals discontinuities in the CA1-subiculum of the hippocampus in freely moving rats. Neuroscience (in press)

Graubard K, Raper JA, Hartline DK (1983) Graded synaptic transmission between identified spiking neurons. J Neurophysiol 50:508–521

Havstad JW, Ehlers CL (1989) Attractor dimension of nonstationary dynamical systems from small data sets. Phys Rev A 39:845–853

Hayashi H, Ishizuka S (1987) Chaos in molluscan neuron. In: Degn H, Holden AV, Olsen LF (eds) Chaos in biological systems. Plenum, New York, pp 157–166

Kaplan DT (1989) The dynamics of cardiac electrical instability. PhD thesis, Harvard University, Cambridge

Labos E (1987) Chaos and neural networks. In: Degn H, Holden AV, Olsen LF (eds) Chaos in biological systems. Plenum, New York, pp 195–206

Mees AI, Rapp PE, Jennings LS (1987) Singular-value decomposition and embedding dimension. Phys Rev A 36:340–346

Rapp PE, Zimmerman ID, Albano AM, de Guzman GC, Greenbaun NN, Bashore TR (1985) Experimental studies of chaotic neural behavior: cellular activity and electroencephalographic signals. In: Othmer HG (ed) Nonlinear oscillations in biology and chemistry. Springer, Berlin Heidelberg New York, pp 175–205

Rapp PE, Albano AM, Mees AI (1988) Calculation of correlation dimensions from experimental data: progress and problems. In: Kelso JAS, Mandell AJ, Shlesinger MF (eds) Dynamic patterns in complex systems. World Scientific Publishers, Singapore, pp 191–205

Rapp PE, Bashore TR, Martinerie JM, Albano AM, Mees AI (1990) Dynamics of brain electrical activity. Brain Topography (in press)

Röschke J (1986) Eine Analyse der Nichtlinearen EEG-Dynamik. Thesis, Universität Göttingen,, pp 1–128

Theiler J (1986) Spurious dimension from correlation algorithms applied to limited time-series data. Phys Rev A 34:2427–2432

Theiler J (1990) Estimating fractal dimension. J Opt Soc Am A (in press)

Chaotic Dynamics in Brain Activity*

A. BABLOYANTZ

1 Introduction

The aim of this paper is to report on a new attempt at characterizing the electro-encephalogram (EEG), which is based on recent progress in the theory of nonlinear dynamical systems (Brandstäter et al. 1983; Nicolis and Nicolis 1984, 1986; Babloyantz et al. 1985). The method is independent of any modeling of brain activity. It relies solely on the analysis of data obtained from a single-variable time series. From such a "one-dimensional" view of the system, one reconstructs the $\{X_k\}$ (where $k = 1, \ldots,$ n) variables necessary for the description of systems dynamics. With the help of these variables, phase-space trajectories are drawn. Provided that the dynamics of the system can be reduced to a set of deterministic laws, the system reaches in time a state of permanent regime. This fact is reflected by the convergence of families of phase trajectories toward a subset of the phase space. This invariant subset is called an "attractor."

Thus, from an analysis of the EEG considered as a time series, it is possible to answer the following questions:

1. Is it possible to identify attractors for various stages of brain activity? In other words, can the salient features of neuronal activities be described by deterministic dynamics?
2. If attractors exist, what is their Hausdorff dimension D? This quantity gives a means of classifying attractors, and the dynamics they portray, as periodic, quasi-periodic, or chaotic.
3. What is the minimum number of variables necessary for the description of a given EEG activity?

2 Phase Portraits

Let us assume that the dynamics of the brain activity is described by a set of $\{X_0(t),$ $X_1(t), \ldots, X_{n-1}(t)\}$ variables satisfying a system of first-order differential equations. A differential equation of order n with a single variable X_0, accessible from experimental data, is equivalent to the original set. Now both X_0 and its derivatives, and therefore the ensemble of n variables, can be obtained from a single time series. However, it is more convenient to construct another set of variables $\{X_0(t),$

---

* Originally published in Başar E (ed) Dynamics of sensory and cognitive processing by the brain. Springer, Berlin Heidelberg New York, pp 196–202 (Springer series in brain dynamics, vol 1). Cross references refer to that volume.

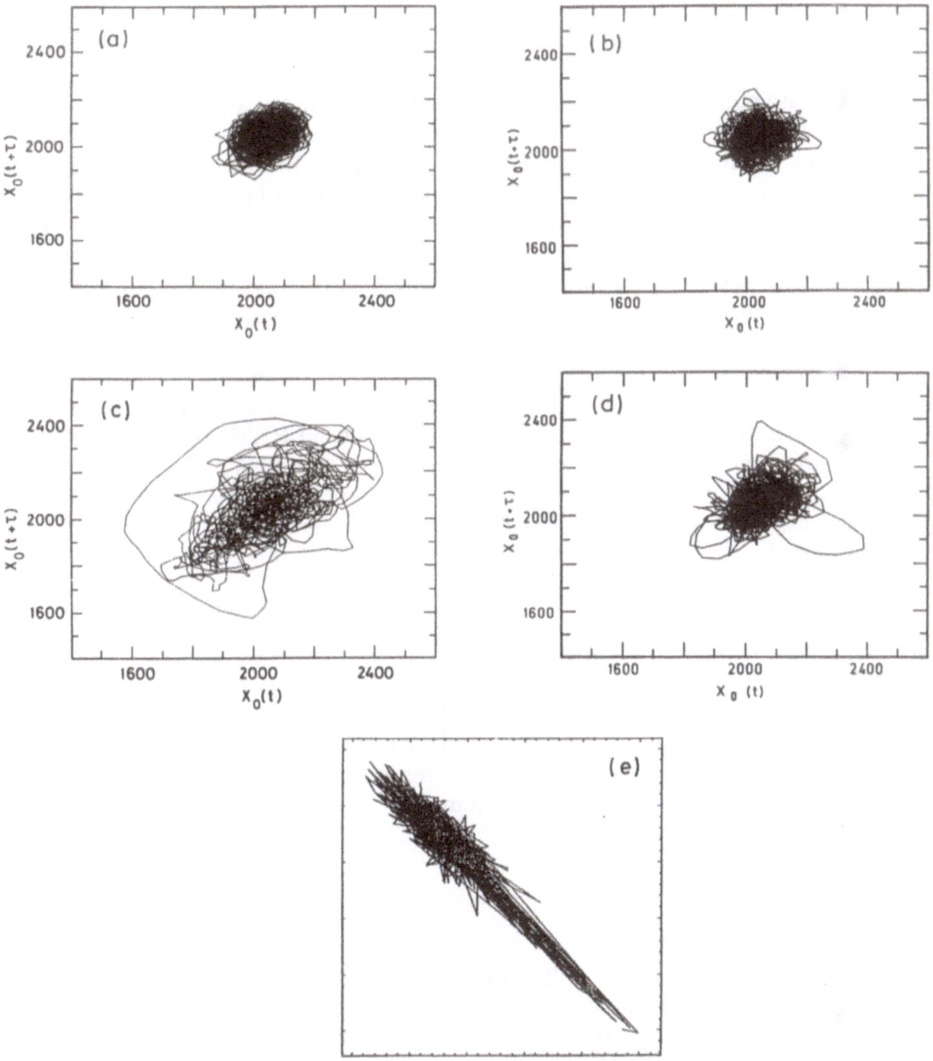

Fig. 1a–e. Two-dimensional phase portraits derived from the EEG of (a) an awake subject, (b) stage 2 sleep, (c) stage 4 sleep, (d, e) REM sleep. The time series $X_0(t)$ is made of $N = 4000$ equidistant points. The central EEG derivation C4-A1 according to the Jasper system was recorded with a PDP 11-44, 100 Hz for 40 s. The value of the shift from 1a to 1e is $\tau = 10\Delta t$

$X_0(t + \tau), \ldots, X_0[t + (n-1)\tau]\}$, which is topologically equivalent to the original set (Takens 1981). X may represent the electrical potential V recorded by the EEG. These variables are obtained by shifting the original time series by a fixed time lag τ ($\tau = m\Delta t$, where m is an integer and Δt is the interval between successive samplings).

These variables span a phase space, which allows the drawing of the phase portrait of the system or, more precisely, its projection into a low-dimensional subspace of the full phase space. With the help of the procedures cited above, we have constructed the phase-space portraits of various stages of sleep cycles.

The phase portrait of the awake subject is densely filled and occupies only a small portion of the phase space (Fig. 1a). The representative point undergoes deviations from some mean position in practically all directions. In sleep stage 2, already a larger portion of the phase space is visited and a tendency toward a privileged direction is seen (Fig. 1b). This tendency is amplified in sleep stage 4 and one sees preferential pathways, suggesting the existence of reproducible relationships between instantaneous values of the pertinent variables (Fig. 1c). This phase portrait is the largest and it exhibits a maximum "coherence," which diminishes again when rapid eye movement (REM) sleep sets in (Fig. 1d).

A universal attractor for different REM episodes of a single night and a given individual seems unlikely, as the REM episodes are associated with intense brain activity and generation of dreams. Figure 1e shows a second REM episode in the sleep cycle of the same individual whose EEG recording was used in Fig. 1d.

We must now determine whether these phase portraits represent chaotic attractors, in which case they could be characterized further by a number corresponding to their dimensionality as defined below.

3 The EEG Attractors

Let us consider an ensemble of points in an n-dimensional phase space. We cover the set with hypercubes of size ε. If $N(\varepsilon)$ is the minimum number of hypercubes necessary to cover the set, the Hausdorff, dimension D of the attractor is defined as (Bergé et al. 1984):

$$D = \frac{\ln N(\varepsilon)}{\ln (1/\varepsilon)} \tag{1}$$

where for small ε, $N(\varepsilon) \simeq \varepsilon^{-D}$.

From this definition, one easily verifies that if the set is reduced to a single point, then $D = 0$. If the set of points represents a segment of line length L, then $N(\varepsilon) = L\varepsilon^{-1}$ and therefore $D = 1$. In the case of a surface S, $N(\varepsilon) = S\varepsilon^{-2}$ and $D = 2$. For these simple cases, the Hausdorff dimension coincides with the euclidean dimension. However, this is not so for a class of objects called "fractals," which may be illustrated by the following example. Let us consider a segment of unit length and remove the middle third of the segment. We repeat the same operation on the remaining segments. If the deletion is performed an infinite number of times, we obtain an infinite number of disconnected points called a "Cantor set." A simple calculation based on Eq. 1 shows that the Hausdorff dimension of this set is $D = 0.63$, which is between 0 and 1. This number is the fractal dimension of the set.

The phase portraits of Fig. 1 belong to the family of fractal objects. However, the Hausdorff dimension of the attractor cannot be evaluated in a simple way from Eq. 1.

Chaotic attractors constructed from a time series can be characterized by another method proposed by Grassberger and Procaccia (1983a, b). Let $\{(X_0(t_1),\ \ldots,$

$X_0[t + (n-1)]\}$ represent the coordinates of a point $\vec{X}$ in the phase space of Fig. 1. Given an $\vec{X}_i$, we compute all distances $|\vec{X}_i - \vec{X}_j|$ from the $N-1$ remaining points of the data. This allows us to count the data points that are within a prescribed distance r from point $\vec{X}_i$ in the phase space. Repeating the process for all values of i, one arrives at the quantity

$$C(r) = \frac{1}{N^2} \sum_{i \neq j=1}^{N} \theta(r - |\vec{X}_i - \vec{X}_j|), \tag{2}$$

where θ is the Heaviside function, $\theta(X) = 0$ if $X < 0$, and $\theta(X) = 1$ if $X > 0$.

The nonvanishing of $C(r)$ measures the extent to which the presence of a data point $\vec{X}_i$ affects the position of the other points. $C(r)$ may thus be referred to as the integral *correlation function* of the attractor.

Let us fix a small ε and use it as a yardstick for probing the structure of the attractor. If the latter is a line, clearly the number of data points within a distance r from a prescribed point should be proportional to r/ε. If the attractor is a surface, this number should be proportional to $(r/\varepsilon)^2$ and, more generally, if it is a d-dimensional manifold, the number should be proportional to $(r/\varepsilon)^d$. We therefore expect that for relatively small r, $C(r)$ should vary as

$$C(r) \sim r^d.$$

In other words, the dimensionality d of the attractor is given by the slope of $\log C(r)$ versus $\log r$ in a certain range of values of r:

$$\log C(r) = d \left| \log r \right| + C^\circ \tag{3}$$

The results cited above suggest the following algorithm:

1. Starting from a time series provided by the EEG, construct the correlation function, Eq. 2, by considering successively higher values of the dimensionality of the phase space.
2. Deduce the slope d near the origin according to Eq. 3 and see how the result changes as n is increased.
3. If the d versus n dependence is saturated beyond some relatively small n, the system represented by the time series should possess an attractor. The saturation value d is regarded as the dimensionality of the attractor represented by the time series. The value of n beyond which saturation is observed provides the minimum number of variables necessary to model the behavior represented by the attractor.

The procedure cited above has been applied to two sets of EEG data corresponding to stage 2 sleep of two individuals and stage 4 sleep of three individuals (Babloyantz et al. 1985). Figure 2 gives the $\log C$ versus $\log r$ dependence for $n = 2$ to $n = 7$ computed for stage 4 sleep. We observe the existence of a region over which this dependence is linear.

The slope of the curve $\log C(r)$ versus $\log r$ has been evaluated with extreme care. After determining the boundaries of the linear zone by visual inspection, we determine the slope of m first points in this segment by using the least-square method. The operation is repeated all along the linear region by sliding m one point further. The computation is repeated for increasing values of m. If the region is linear, all these

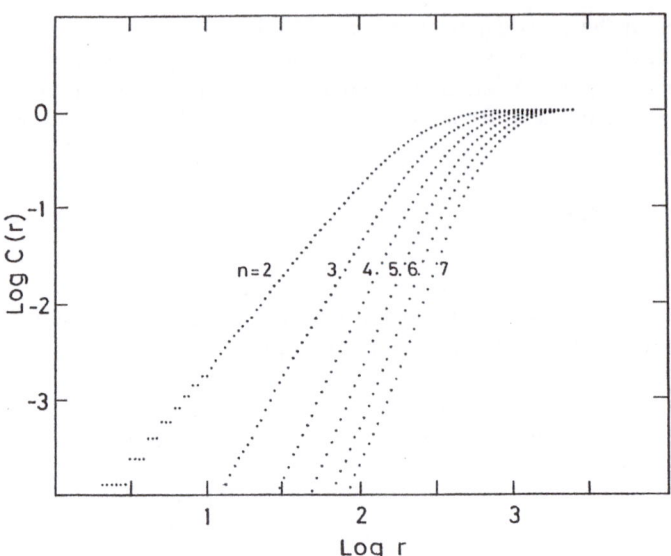

Fig. 2. Dependence of the integral correlation function $C(r)$ on the distance r for stage 4 sleep. Parameter values as in Fig. 1

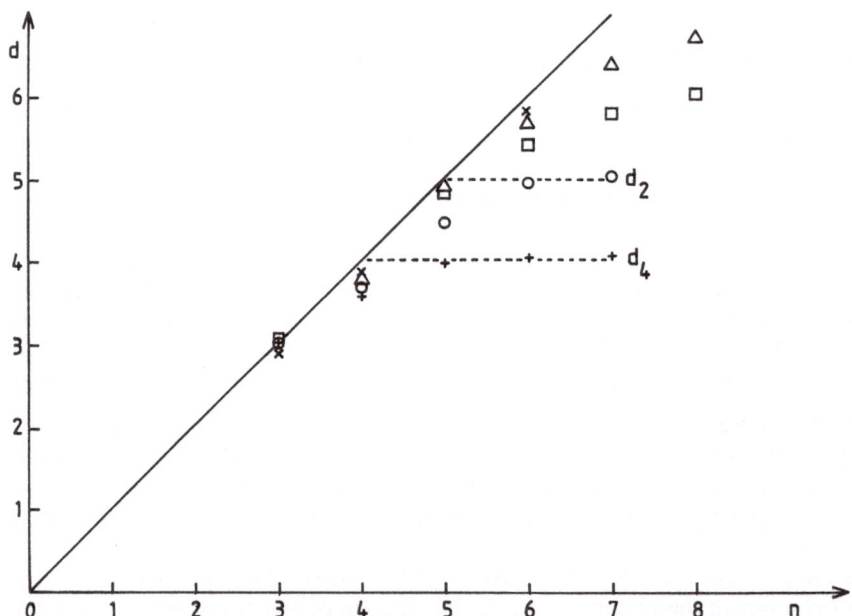

Fig. 3. Dependence of dimensionality d on the number of phase space variables n for a white noise signal (x), the EEG attractor of an awake subject ($\triangle$), stage 2 sleep (o), stage 4 sleep ($+$), and REM sleep ($\square$), for the same number of data points as in Fig. 1

operations must yield the same value of the slope (within acceptable error bound-aries).

Although in principle every value of time lag τ is acceptable for the resurrection of the system's dynamics, in practice, for a given time series, only a well-defined range of τ gives satisfactory results.

Figure 3 shows saturation curves describing the dependence of the slope d versus the dimension n of the embedding phase space computed for the awake state and for several stages of the sleep cycle. They are compared with the behavior obtained from a random process such as Gaussian white noise. There is no saturation for the awake state; however, we are far from random noise. A satisfactory saturation exists for stage 2 sleep. In this case, we find $d = 5.03 + 0.07$ and $d = 5.05 + 0.1$. Saturation curves for stage 4 sleep show $d = 4.08 \pm 0.05$, $d = 4.05 + 0.05$, and $d = 4.37 + 0.1$. For the REM state, saturation is again poor.

4 Time-Dependent Property

The chaotic nature of the attractor can also be assessed with the help of a time-dependent property. Although in the presence of a chaotic attractor all trajectories converge toward a subset of the phase space, inside the attractor, two neighboring trajectories may diverge. This fact reflects the extreme sensitivity of chaotic dynamics to the initial conditions. The rate of the divergence of the trajectories in time may be assessed from a time series (Wolf et al. 1985). The Lyapunov exponents λ_i are the average of these individual evaluations over a large number of trials. A negative Lyapunov exponent indicates an exponential approach of the initial conditions on the attractor; on the contrary, a positive λ_i expresses the exponential divergence on an otherwise stable attractor. Thus, a positive Lyapunov exponent indicates the presence of chaotic dynamics.

Using the Fortran code described by Wolf et al. (1985), we evaluated the largest positive Lyapunov exponent λ for stage 2 and stage 4 of deep sleep. For stage 2, we find a positive value of λ_2 between 0.4 and 0.8. For stage 4, we find also a positive number $0.3 < \lambda_4 < 0.6$. The inverse of this quantity gives the limits of predictability of the long-term behavior of the system.

5 Conclusions

We have shown that from a routine EEG recording, the dynamics of brain activity could be reconstructed. The fact that chaotic attractors could be identified for two stages of normal brain activity indicates the presence of deterministic dynamics of a complex nature. This property should be related to the ability of the brain to generate and process information.

Unlike periodic phenomena, which are characterized by a limited number of frequencies, chaotic dynamics show a broadband spectrum. Thus, chaotic dynamics increase the resonance capacity of the brain. In other words, although globally a chaotic attractor shows asymptotic stability, there is an internal instability reflected by the presence of positive Lyapunov exponents. This results in a great sensitivity to the initial conditions and, thus, an extremely rich response to external input.

The topological properties of the attractors and their quantification by means of dimensionality analysis may be an appropriate tool in the classification of brain activity and, thus, a possible diagnostic tool. For example, various forms of epileptic seizures could be classified according to their degree of coherence (Babloyantz and Destexhe 1986). Moreover, the determination of the minimum number of variables necessary for the description of epileptic attractors is a valuable clue for model construction.

References

Babloyantz A, Destexhe A (1986) Low-dimensional chaos in an instance of epilepsy. Proc Natl Acad Sci USA 83:3513–3517

Babloyantz A, Nicolis C, Salazar M (1985) Evidence of chaotic dynamics of brain activity during the sleep cycle. Phys Lett [A] 111:152–156

Berge P, Pomeau Y, Vidal C (1984) L'ordre dans le chaos: vers une approche déterministe de la turbulence. Hermann, Paris

Brandstäter A, Swift J, Swinney HL, Wolf A (1983) Low-dimensional chaos in a hydrodynamic system. Phys Rev Lett 51:1442–1445

Grassberger P, Procaccia I (1983a) Characterization of strange attractors. Phys Rev Lett 50:346–349

Grassberger P, Procaccia I (1983b) Measuring the strangeness of strange attractors. Physica [D] 9:189–208

Nicolis C, Nicolis G (1984) Is there a climatic attractor? Nature 311:529–532

Nicolis C, Nicolis G (1986) Reconstruction of the dynamics of the climatic system from time-series data. Proc Natl Acad Sci USA 83:536–540

Takens F (1981) Detecting strange attractors in turbulence. In: Rand DA, Young LS (eds) Dynamical systems and turbulence, Warwick 1980. Springer, Berlin Heidelberg New York pp 366–381 (Lecture notes in mathematics, vol 898)

Wolf A, Swift JB, Swinney HL, Vastano JA (1985) Determining Lyapunov exponents from a time series. Physica [D] 16:285–317

The EEG is Not a Simple Noise: Strange Attractors in Intracranial Structures*

J. Röschke and E. Başar

1 Introduction

Since Berger's (1929) first description of the electrical activity of the brain, several approaches have been undertaken in order to correlate the activity at neuronal levels with the origin of the electroencephalogram (EEG). Creutzfeldt (1974) pointed out that the spontaneous electrical activity of the CNS and sensory evoked potentials are highly correlated to intracellularly measured postsynaptic potentials (EPSPs and IPSPs). Ramos et al. (1976) postulated that it is impossible to specify any general causal or predictable relationship between the waveform of an evoked potential and the firing pattern of a neuron. Some authors take the view that the spontaneous EEG activity is an expression of the incessant, irregular background neural firing. Do we have the right to consider the spontaneous activity of the brain as a background noise in the sense of ideal communication theory? Or rather, is the EEG a most important fluctuation, which controls the sensory evoked and event-related potentials? We have written elsewhere that the spontaneous activity plays an active role in the signals transmitted through various structure and recorded at various sites in the brain and that the EEG should not be considered as a noisy signal. Especially, we have assumed that regular patterns of the EEG reflect coherent states of the brain during which cognitive and sensory inputs are processed (Başar 1980, 1983a, b).

The main goal of the present paper is to show that the brain's spontaneous activity is not a simple noise, but is an active signal probably reflecting causal responses from hidden events and sources during sensory and cognitive processing in the brain.

There are several difficulties to overcome in order to describe well-defined states of spontaneous activity and evoked potentials. Although conventional methods of system theory, such as power spectral analysis, have been very useful for analyzing the brain waves as a first approach, the highly nonlinear character of the brain's dynamic behavior led us first to use phase portrait analysis, analogies with laser theory, and the Duffing equation (Başar 1980, 1983a, b). Our preliminary results from an analysis of the EEG in phase space allowed us to speculate on the existence of strange attractors in the EEG (Başar 1983a, b).

In our newest approach we used the algorithm of Grassberger and Procaccia (1983), similar to the analysis of Babloyantz and Nicolis (1985). The EEG signal was embedded into phase space and we computed the dimension of the attractors of the acoustical cortex (GEA), the hippocampus (HI), and the reticular formation (RF) of the cat brain during slow-wave sleep stage (SWS).

---

* Originally published in Başar E (ed) Dynamics of sensory and cognitive processing by the brain. Springer, Berlin Heidelberg New York, pp 203–216 (Springer series in brain dynamics, vol 1). Cross references refer to that volume.

Our preliminary results showed that the field potentials (or EEG) in intracranial structures do not reflect the behavior of a simple noisy signal, as has been shown for the human EEG derived from scalp electrodes (Babloyantz and Nicolis 1985). However, field potentials from various brain structures have properties of strange attractors, indicating the presence of a chaotic system. The most important result in the present paper is the existence of differentiated dimensionality in functionally independent brain structures. This in turn makes the future application of this method most useful for the differentiated analysis of brain states and function. In this study we also discuss what is meant by "attractors" and "strange attractors", and our belief that the use of such concepts in EEG research may lead to basic trends (see also the Epilogue).

2 The Mathematical Procedure and the Concept of "Attractor"

In order to describe periodic, aperiodic, or even chaotic behavior of nonlinear systems arbitrarily with more degrees of freedom, several approaches have been applied. Lorenz (1963) applied concepts of nonlinear dynamics to the convection phenomenon of hydrodynamics in order to describe atmospheric turbulence (Navier-Stokes equation). He demonstrated the possibility that the unpredictable or chaotic behavior observed in such an infinite-dimensional system might be caused by a three-dimensional (deterministic) dynamical system.

In order to understand these arguments, we have to consider some recent tools from the theory of nonlinear dynamical systems. The description of systems behavior (in our case the EEG from different brain structures) must be analyzed not only in the time domain or frequency domain, but also in the *phase space*. In general, a phase space is identified with a topological manifold. An n-dimensional phase space is spanned by a set of n independent linear vectors. This requirement is generally sufficient. There are several possibilities for defining a phase space. We consider a proposal of Takens (1981) and span a ten-dimensional phase space by $x(t), x(t+\tau)$, $\ldots, x(t+9\tau)$, where τ means a fixed time increment. Every instantaneous state of a system is therefore represented by a set $(x_1, \ldots, x_n)$, which defines a point in the phase space. The sequence of such states (or points) over the time scale defines a curve in the phase space, called a "trajectory". As time increases, the trajectories either penetrate the entire phase space or they converge to a lower-dimensional subset. In this latter case, the set to which the trajectories converge is called an "attractor". Figure 1 shows some simple (converging) attractors and a noise that does not converge in a two-dimensional phase space.

In relation to the topological dimension of the remaining attractor, one can deduce various properties of the investigated system. If the dimension of an *attractor* is a noninteger, called a "fractal", the attractor is a "strange attractor" and can be identified with the properties of deterministic chaos. A characteristic phenomenon of deterministic chaos is a sensitive dependence on initial conditions. Similar causes do not produce similar effects. This is a very extensive statement, which apparently damages the causality principle of natural philosophy. However, by examining the properties of a strange attractor more precisely, one finds that a strange attractor

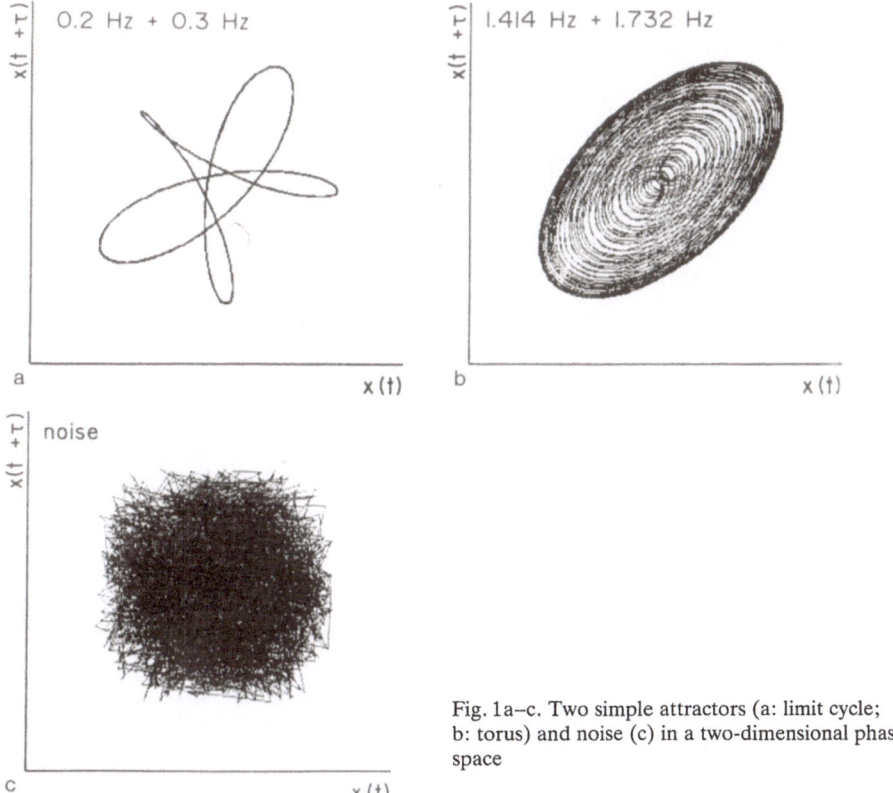

Fig. 1a–c. Two simple attractors (a: limit cycle; b: torus) and noise (c) in a two-dimensional phase space

may have a strong conformity, called „self-similarity", which is an invariance with respect to scaling.

Self-similar objects possess a fractal dimension (Schroeder 1986). What is a "fractal dimension"? One of the oldest notions of dimension is that of a topological dimension D_T. For a point, $D_T = 0$; for a line, $D_T = 1$, and for a plane, $D_T = 2$. A first generalization is the *Hausdorff dimension or fractal dimension* D_F. For simple sets, for example a limit cycle or a torus, the fractal dimension D_F is an integer and is equal to the topological dimension D_T. For a n-dimensional phase space, let $N(\varepsilon)$ be the number of n-dimensional balls (or cubes) of radius ε required to cover an attractor. Then the fractal dimension D_F is defined as

$$D_F = \lim_{\varepsilon \to 0} \frac{\log N(\varepsilon)}{|\log \varepsilon|}$$

The classical example of a set whose fractal dimension exceeds its topological dimension [such sets are called "fractals" by Mandelbrot (1977)] is Cantor's set (Fig. 2). If one chooses $\varepsilon = \left(\frac{1}{3}\right)^n$, then $N(\varepsilon) = 2^n$, and it follows that

$$D_F = \lim_{\varepsilon \to 0} \frac{\log N(\varepsilon)}{|\log \varepsilon|} = \lim_{n \to \infty} \frac{\log 2^n}{\log 3^n} = \frac{\log 2}{\log 3} = 0.630 \ldots$$

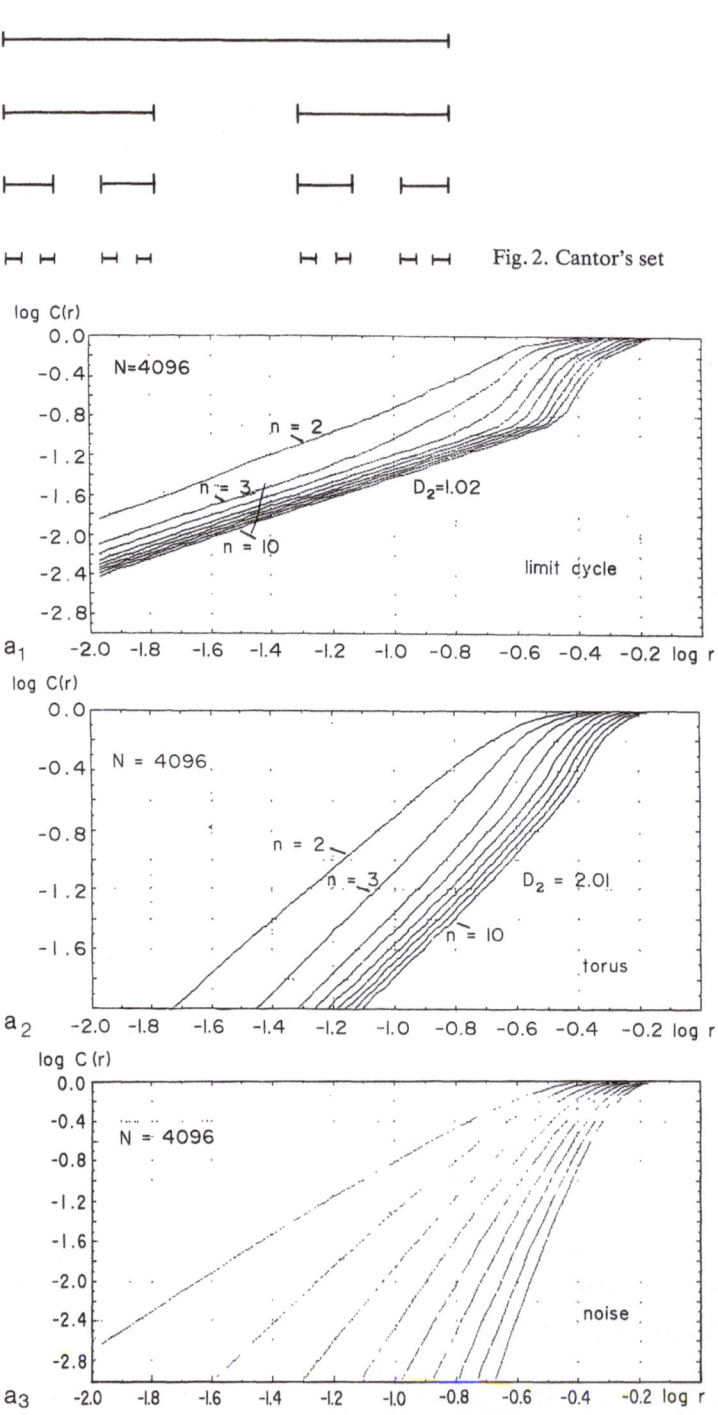

Fig. 2. Cantor's set

Fig. 3a. Log $C(r)$ versus log r for the data shown in Fig. 1

Note the self-similarity of Cantor's set. For example, the fractal dimension of the Lorenz attractor is $D_F = 2.03$. A generalization of the fractal dimension is introduced in information theory. The Renyi information of order q is defined as

$$I_q = \frac{1}{1-q} \log \sum_{i=1}^{N(R)} p_i^q \qquad (q \neq 1)$$

$$I_q = - \sum_{i=1}^{N(R)} p_i \log p_i \qquad (q = 1)$$

Let p_i be the probability that an arbitrary point (of an attractor) falls into cube i with-radius R and let $N(R)$ be the number of nonempty cubes. The generalized dimensions D_q of order q are given by

$$D_q = \lim_{R \to 0} \frac{I_q(R)}{\log(1/R)}$$

For $q = 0$ we find $D_0 = D_F$. D_1 is called the "information dimension" and D_2 is called the "correlation dimension". It is the case that

$$D_0 \geq D_1 \geq D_2 \geq \ldots$$

In practice, the correlation dimension D_2 is the generalized dimension easiest to estimate from attractors generated by experimental data (Grassberger 1984, personal communication), because

$$I_2 = - \log \sum_{i=1}^{N(R)} p_i^2 = - \log C(R)$$

where $C(R)$ is a measure of the probability that two arbitrary points $\tilde{x}$, $\tilde{y}$ will be separated by distance R. $C(R)$ is called the "correlation integral" and can be easily computed:

$$C(R) = \lim_{N \to \infty} \frac{1}{N^2} \sum_{\substack{i,j=1 \\ i \neq j}}^{N} \theta(R - |\tilde{x} - \tilde{y}|)$$

where θ is the Heavyside function.

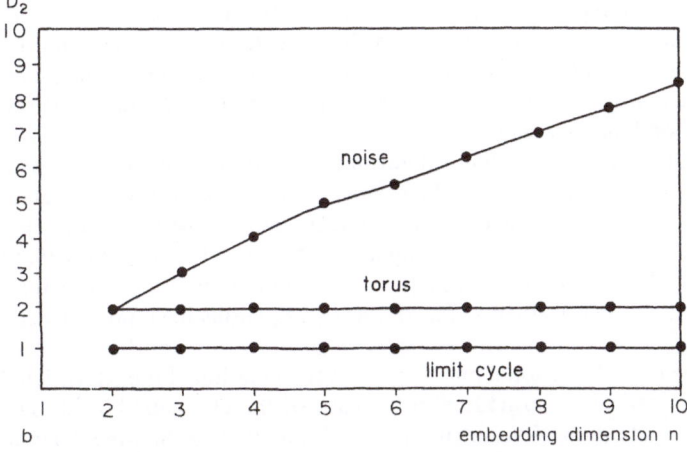

Fig. 3b. The correlation dimension D_2 versus the embedding dimension n for the limit cycle, the torus, and noise shown in Fig. 1

It then follows that

$$D_2 = \lim_{R \to 0} \frac{\log C(R)}{\log R}$$

or

$$C(R) \sim R^{D_2}$$

The main point is that $C(R)$ behaves as a power of R for small R. This means that it is possible to find a measure for the dimensionality of an attractor by evaluating $C(R)$ and plotting $\log C(R)$ versus $\log R$. For the Lorenz attractor, Grassberger and Procaccia (1983) found $D_2 = 2.05$.

Figure 3 shows the computation of the correlation dimension for the examples presented in Fig. 1. It is obvious that a noise signal has no attractor. By plotting the slope of the curves versus the embedding dimension, no saturation can be observed. On the contrary, there exist attractors of $D_2 = 1.00$ for a limit cycle and $D_2 = 2.00$ for quasi-periodic flow.

3   Experimental Procedure and Results

In order to analyze the dimensionality of field potentials, five cats with chronically implanted electrodes were studied. The chronic electrodes were implanted in the GEA, HI, and RF. In total, 15 experimental trials during SWS activity were evaluated. The intracranial EEG signals were digitized by a 12-bit AD converter and stored in the memory of an HP 1000-F computer. The sampling frequency was $f_s = 100$ Hz for all trials.

Dimensions of the EEG signals were evaluated over a time period of about 20 s ($N = 2048$) and 40 s ($N = 4096$). Details of the software have been described elsewhere (Röschke 1986). The phase space was constructed by using the time-delayed coordinates proposed by Takens. Theoretically, the evaluation of the dimension of the attractors should be independent of the arbitrary but fixed time increment τ. In practice, this independence is not generally valid. Investigations of low-pass filtered noise (non-deterministic signals) have shown that dD_2/dn depends on the time increment τ. It is evident that in the case of nondeterministic signals, $dD_2/dn \neq 0$ is observed for every embedding dimension.

However, by evaluating signals from a deterministic system, it is observed as a rule that the dimension D_2 of an attractor does converge towards a saturation value. In this case, $dD_2/dn = 0$ and this convergence is independent of the choice of τ in a given interval $\tau_1 < \tau < \tau_2$. Some recent investigations (Fraser 1985; Holzfuss 1985) have assumed that the best choice of τ corresponds to a minimum of the "mutual information" between two measurements, but these investigations have not yet been properly concluded.

Figure 4 shows the two-dimensional phase space representation of the EEG signal for the three investigated brain structures in the case of the cat named Toni. The time period used for evaluation of these curves was about 40 s and the time delay to construct the phase space was about 60 ms. Figure 5 shows the plot of $\log C(r)$ versus $\log r$. In all the computations, the EEG signal was embedded in a ten-dimensional

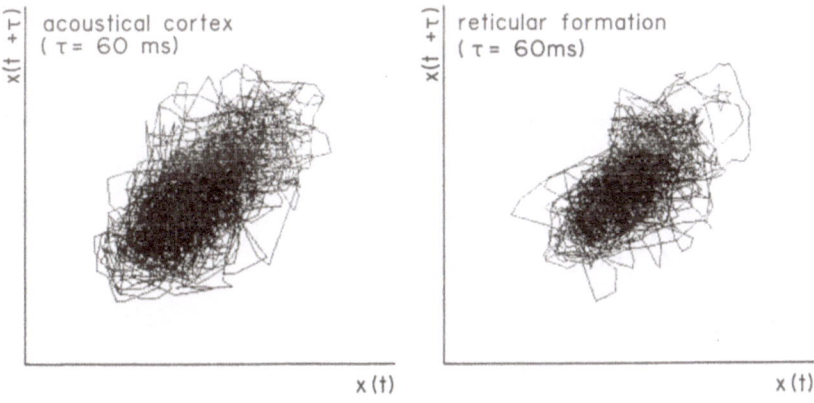

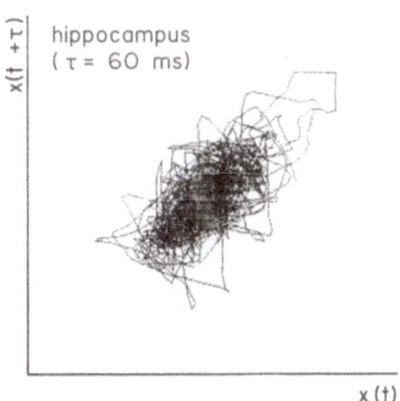

Fig. 4. Two-dimensional phase space representation of the EEG attractors from the brain structures investigated

phase space. Figure 6 shows the convergence of D_2 (slope of the curves from Fig. 5) towards the saturation value as a function of the embedding dimension n. Especially in this case, the dimensions of the attractors had the following values:

$$D_{GEA} = 5.00 \pm 0.10$$
$$D_{RF} = 4.25 \pm 0.07$$
$$D_{HI} = 4.32 \pm 0.07$$

Table 1 presents the correlation dimension D_2 computed for all of the experimental data from five cats and 15 experiments.

1. For both $N = 2048$ and $N = 4096$, one cannot determine an unambiguous conformity. Both the results from a single cat and the dimensions from a single region vary within the range of acceptable limits.
2. In 86% of the investigated trials, D_{GEA} presents the maximal dimension. By taking into account all the evaluated data, the following mean values have been obtained:

$$D_{GEA} = 5.06 \pm 0.31$$
$$D_{RF} = 4.58 \pm 0.38$$
$$D_{HI} = 4.37 \pm 0.36$$

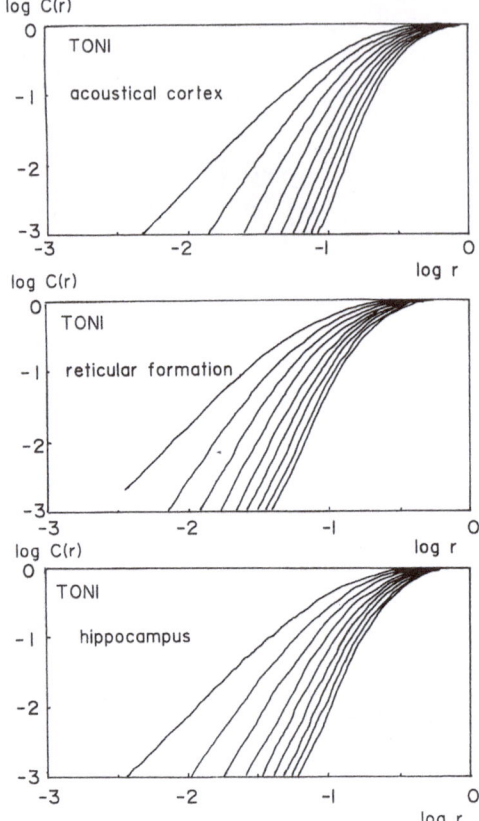

Fig. 5. Log $C(r)$ versus log r for the cat named Toni

In other words, the data confirm the following important relation:

$$D_{GEA} > D_{RF} > D_{HI}$$

4   Discussion

The concept of dynamic patterns to use in understanding bioelectric phenomena was proposed by Katchalsky et al. (1974). A fundamental problem in the physical as well as in the biological sciences is the origin of a dynamic pattern. In physical science this problem can be attacked at vulnerable points; i.e., in systems that are simple enough to permit analysis both in physical and mathematical terms, dynamic patterns refer to those patterns that arise and are maintained by the dissipation or consumption of energy, such as *traveling or standing waves* generated in the air, on the surface of water, or by a vibrating violin string. They can be contrasted to static patterns, such as a stack of nesting chains, a crystal, or a virus capsid.

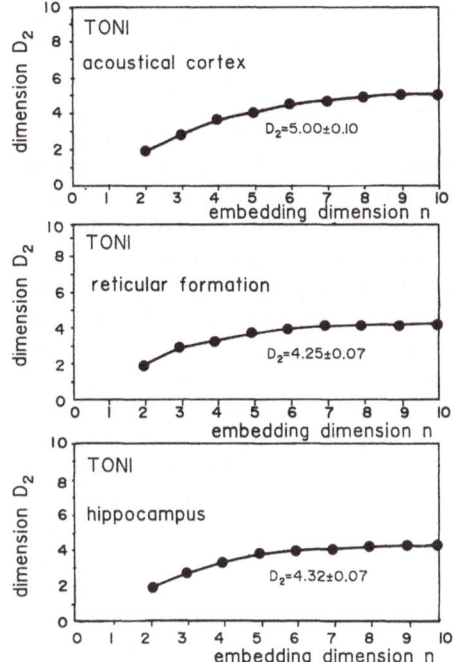

Fig. 6. The convergence of the slope D_2 of the curves (from Fig. 5) towards the saturation value as a function of the embedding dimension n

Table 1. Correlation dimension D_2 computed for all of the experimental data from five cats and 15 experiments

$N = 2048$	TONI	LENA	DESY	SARA	ROMY	LENA	SARA	D_2
GEA	5.00	4.35	4.66	4.93	5.00	5.00	–	4.82 ± 0.27
RF	4.25	4.62	4.30	4.24	5.22	4.60	4.05	4.48 ± 0.39
HI	4.32	4.35	4.00	3.80	5.00	4.15	4.85	4.35 ± 0.44

$N = 4096$	TONI	LENA	ROMY	DESY	SARA	TONI	LENA	ROMY	D_2
GEA	5.30	5.00	5.17	5.30	5.10	5.00	5.60	5.45	5.24 ± 0.21
RF	4.20	4.35	4.75	5.15	4.76	4.20	5.00	5.00	4.68 ± 0.38
HI	4.40	4.28	4.44	4.66	4.00	3.90	4.61	4.76	4.38 ± 0.31

According to Katchalsky et al. (1974), the central question is: how does uniform matter, obeying physical principles, i.e., laws of conservation of momentum, matter, and energy, spontaneously develop regular patterns? In other words, how is it that a set of isotropic causes can give rise to anisotropic dynamic effects? This appears to be the root problem of morphogenesis; extending from it are the more widely encountered problems of how preexisting static structures influence dynamic patterns.

In the book by Katchalsky et al. (1974), some dynamic patterns observed in geology, meteorology, and astrophysics are also described; for example, dynamic patterns on a large scale in clouds and the solar coronasphere. According to traditional Newtonian mechanics, if certain things are known about a system – all of the forces acting on it, its position, and the velocity of its particles – it is possible to describe, in theory, all of its future states. However, the current that started with the pioneering work of H. Poincaré at the turn of this century has made clear that the predictability of even classical deterministic systems can be quite limited. Simple nonlinear systems which are just as deterministic as the motions of the planets can behave in a manner so erratic as to prelude predictability past a short time (Shaw 1981). The existence of these "chaotic" systems raises both practical and conceptual questions.

A simple example is described by Hooper (1983): "Suppose you are sitting beside a waterfall watching a cascade of white water flow regularly over jagged rocks, when suddenly a jet of cold water splashes you in the face. The rocks have not moved, nothing has disrupted the water, and presumably no evil sprites inhibit the waterfall. So why does the water suddenly "decide" to splash you?" Physicists studying fluid turbulence have wondered about this kind of thing for several hundred years, and only recently have they arrived at some conclusions that seem to solve the problem at least in part: the waterfall's sudden random splashes do not come from some inperceptible jiggle, but from the inner dynamics of the system itself. Behind the chaotic flow of turbulent fluids or the shifting cloud formations that shape the weather lies an abstract descriptor which the physicists call a "strange attractor." What is an "attractor" and what makes it "strange"? We shall try to describe it again by using the simple explanation of Hooper (1983). Suppose one puts water in a pan, shakes it up, and then stops shaking it; after a time it will stop whirling and come to rest. The state of rest – the equilibrium state – can be described mathematically as a "fixed point," which is the simplest kind of attractor.

Let us now imagine the periodic movement of a metronome or a pendulum swinging from left to right and back again. From the viewpoint of geometry, this motion is said to remain within a fixed cycle forever. This is the second kind of attractor, the *limit cycle*. All of the various types of limit cycles share one important characteristic: *regular, predictable motion*. The third variety, the *strange attractor*, is *irregular, unpredictable*, or simply *strange*. For example, when a heated or moving fluid moves from a smooth, or laminar, flow to wild turbulence, it switches to a strange attractor.

Chaotic behavior in deterministic systems usually occurs through a *transition* from an orderly state when an external parameter is changed. In studies of these systems, particular attention has been devoted to the question of the route by which the chaotic state is approached. An increasing body of experimental evidence supports the belief that apparently random behavior observed in a wide variety of physical systems is caused by underlying deterministic dynamics of a low-dimensional chaotic (strange) attractor. The behavior exhibited by a chaotic attractor is predictable on short time scales and unpredictable (random) on long time scales.

The unpredictability, and so the attractor's degree of chaos, is effectively measured by the parameter "dimension". Dimension is important to dynamics because it provides a precise way of speaking of the number of independent variables inherent in a motion. For a dissipative dynamical system, trajectories that do not diverge to infinity approach an attractor (Farmer 1982).

The dimension of an attractor may be much less than the dimension of the phase space that it sits in. In other words, once transients die out, the number of independent variables to the motion is much less than the number of independent variables required to specify an arbitrary initial condition. With the help of the concept of dimension it is possible to discuss this precisely. For example, if the attractor is a fixed point, there is no variation in the final space position; the dimension is zero. If the attractor is a limit cycle (pendulum) the phase space varies along a curve; the dimension is one. Similarly, for quasi-periodic motion with n incommensurate frequencies, motion is restricted to an n-dimensional torus (dimension of chaotic attractors).

Noise is a common phenomenon in systems with many degrees of freedom. Under the influence of noise, observables show irregular behavior in the time and broadband Fourier spectra. There is an important difference between a noise signal and chaotic fluctuations resulting from the motion of a larger number of system dimensions. The noise signal does not have a finite dimension, whereas chaotic systems with differential equations show finite dimensionality. This difference can be shown by the evaluation of the dimension.

The field potentials of the cat brain showed almost stable mean values of 5.06, 4.58, 4.37 for the GEA, the HI, and the RF, respectively. In other words, various structures of the brain indicate the existence of various chaotic attractors with fractal dimensions. The signals measured in these different structures do not reflect properties of noise signals, but reflect behavior of strange attractors of quasi-low dimension. The measured dimensions in these various structures are seemingly different attractors which might be functionally uncoupled; in other words, even during the SWS stage, where a state of hypersynchrony is observed in all of the various brain structures, the attractors show significant differences. These differences are stable and statistically relevant. We want especially to report that the dimension of the GEA is significantly higher than that of the HI and the RF. This is a kind of differentiated behavior that cannot be observed by the analysis of power spectra (see Fig. 7). In simple words, the cat GEA seems to show a more complex behavior with a greater degree of freedom than do such structures as the RF and HI. There is *no limit cycle behavior* in the studied structures (see also the Epilogue of this book).

Our computations, which are not yet finished and may be theoretically imperfect, showed that the spontaneous activity of the cat cortex depicted a dimension of around 8–9. The acoustical evoked potentials during the waking stage showed much lower dimensionality than did the spontaneous EEG during the same waking state: the dimension of the evoked potential usually varied between 3.5 and 5.

In our earlier studies, we pointed out the possibility of modeling globally the evoked potentials of the brain with the Duffing equation, assuming that nonlinear forced oscillations of various brain structures could be described with solutions of the Duffing equation or similar equations (Başar 1980, 1983b). We argued further that neural populations of the brain may be regarded as a large number of coupled oscillators, each comprising millions of neurons. If n oscillators are left uncoupled, their attractor will be an n-dimensional torus, with n independent frequencies. If the oscillators are coupled, however, the dimension will be reduced. It is possible, for example, that they will entrain at a single frequency (limit cycle). In a condition of slightly less entrainment, the attractor might have three independent frequencies, i.e., be a three-dimensional torus. One might think that with even less entrainment, the attractor could be a four-dimensional torus. Our preliminary analysis of the brain's

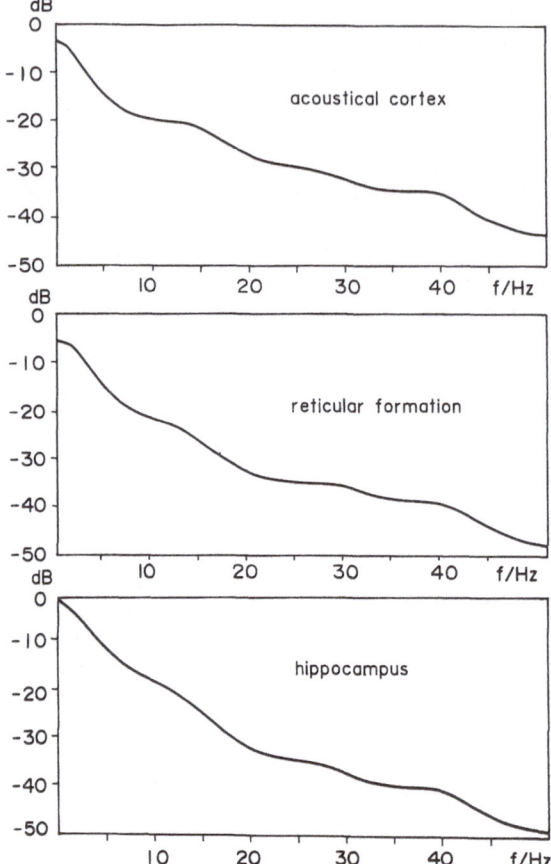

Fig. 7. Power spectra of various structures of the cat brain during the slow-wave sleep stage

spontaneous activity had already pointed out the possibility of correlating the EEG with the behavior of a strange attractor (Başar 1983b). The present results now show that the new algorithm of Grassberger and Procaccia offers an index of the degrees of freedom in the spontaneous activity of the brain. The results presented in this study have a number of new implications.

Three important applications are straightforward:

1. If the fractal dimension is a more precise indicator of state changes of the brain than are power spectra, it can be used in addition to power spectra to describe the state changes in every dimension of EEG analysis: for example, in pharmaco-encephalography and in studies of the evolution of the EEG by comparative analysis of the field potentials of vertebrates, low vertebrates, and invertebrates.
2. Even in the same brain there can exist structurally and functionally independent EEG dimensions. Can one structure of the brain undergo a transition, say from a low-dimensional to a high-dimensional level, while another structure shows the opposite behavior? It is difficult to interpret changes in entropy. The use of the

dimension concept seems to be an adequate method for describing ordered states and for describing dynamic transitions impossible to see in power spectral analysis.

3. It seems that various intracranial structures have various independent attractors. It is possible to describe the ensemble of attractors with a matrix configuration in which various substructures of the brain can occupy a defined place in the matrix. In this case the transition matrices could tell a lot about the multiplicity of attractors as well as about the coupling and decoupling of attractors. The description of multiple attractors could be very useful for the description of cognitive processes, for which sufficiently descriptive parameters are still missing.

The results of the present study constitute a step forward in our preliminary efforts to try to describe the evoked potentials of the brain and the EEG as forced nonlinear oscillations, for which we tried to correlate the processes involved in the Duffing equation. Haken (1985) pointed out interesting formal analogies between laser and brain. We also pointed out similar formal analogies in the generation of evoked potential patterns (Başar 1983b). However, at present, formal analogies with a laser can be considered only for a simple laser. By using fractal dimensions to describe the EEG and the evoked potential, we believe that we can find new descriptors to describe neural analogies with a multimodal laser.

References

Babloyantz A, Nicolis C, Salazar M (1985) Evidence of chaotic dynamics of brain activity during the sleep cycle. Phys Lett [A] 111:152–156

Başar E (1980) EEG-brain dynamics. Relation between EEG and brain evoked potentials. Elsevier/ North-Holland, Amsterdam

Başar E (1983a) Toward a physical approach to integrative physiology. I. Brain dynamics and physical causality. Am J Physiol 245(4):R510–R533

Başar E (1983b) Synergetics of neuronal populations. In: Başar E, Flohr H, Haken H, Mandell AJ (eds) Synergetics of the brain. Springer, Berlin Heidelberg New York

Berger H (1929) Über das Elektroencephalogramm des Menschen. Arch Psychiatr Nervenkr 87: 527–570

Creutzfeldt OD (1974) The neuronal generation of the EEG. In: Renard A (ed) Handbook of electroencephalography and clinical neurophysiology. Elsevier, Amsterdam

Farmer JD (1982) Dimension, fractal, measures, and chaotic dynamics. In: Haken H (ed) Evolution of order and chaos. Springer, Berlin Heidelberg New York

Fraser AM (1985) Using mutual information to estimate metric entropy in dimensions and entropies in chaotic systems. In: Mayer-Kress G (ed) Dimensions and entropies in chaotic systems. Springer, Berlin Heidelberg New York Tokyo

Grassberger P, Procaccia I (1983) Measuring the strangeness of strange attractors. Physica [D] 9:183–208

Holzfuss J (1985) An approach to error-estimation in the application of dimension algorithms. In: Mayer-Kress G (ed) Dimensions and entropies in chaotic systems. Springer, Berlin Heidelberg New York Tokyo

Hooper J (1983) What lurks behind the wild forces of nature? Ask the connoisseurs of chaos. Omni 5:85–92

Katchalsky AK, Rowland W, Blumenthal R (1974) Dynamic patterns of brain cell assemblies. MIT Press, Cambridge

Lorenz EN (1963) Deterministic nonperiodic flow. Atmos. Sci. 20:130

Mandelbrot B (1977) Fractals, form, chance and dimension. Freeman, San Francisco

Ramos A, Schwartz E, John ER (1976) Evoked potential–unit relationship in behaving cats. Brain Res Bull 1:69–75

Röschke J (1986) Eine Analyse der nichtlinearen EEG-Dynamik. Dissertation, University of Göttingen

Schroeder MR (1985) Number theory in science and communications. Springer, Berlin Heidelberg New York Tokyo

Shaw R (1981) Strange attractors, chaotic behaviour, and information flow. Z Naturforsch 36a:80

Takens F (1981) Detecting strange attractors in turbulence. In: Rand A, Young LS (eds) Dynamical systems and turbulence, Warwick 1980. Springer, Berlin Heidelberg New York, pp 366–381 (Lecture notes in mathematics, vol 898)

Nonlinear Neural Dynamics in Olfaction as a Model for Cognition*

W. J. FREEMAN

1 Introduction

The forebrain of primitive vertebrates is so heavily devoted to olfaction that for half a century investigators were misled into considering the function of the hippocampus as being exclusively olfactory. For example, the anterior third of the forebrain of the tiger salamander forms the bulb, the medial third is hippocampus, and the lateral third comprises the piriform and striato-amygdaloid complex (Herrick 1948). According to Herrick, a transitional zone in the mantel receives thalamic axons that convey input to the forebrain from all other sensory systems. He proposed that with the expansion and increasing dominance of these other systems, the brain expanded by adding new parts while preserving the topology of connections of those parts already existing. This view has survived to the present with modifications; it is as if, seeing that olfaction was a success, other systems moved in and co-opted the machinery of the forebrain. Olfaction remains the simplest among the sensory systems. For this reason, if for no other, the study of sensation and cognition might well begin with the sense of smell. But there are three other good reasons: the parallels that exist between olfaction and other senses in their psychophysics, in the dynamics of the masses of neurons comprising them, and in the types of neural activity that they generate.

2 Psychophysics

The olfactory system resembles other sensory systems in consisting of a surface array of receptors of multiple kinds that project in parallel to arrays of central neurons. Some examples of stimuli that are comparable to an odorant are the sight of a constellation such as Orion in the winter sky, the feeling of putting on a coat that is not the correct size, and the sound of a tone that allows immediate identification of the instrument being played — a piano, oboe, etc. These operations are rapid, spatial, and global, and they depend on past experience. The information is expressed by spatial relationships among activated and equally importantly nonactivated receptors, without reference to simple geometric forms. The time frames are longer than that of an action potential but shorter than a heart beat; according to Efron (1970), on the order of 0.1 s.

All of these systems are legendary for their sensitivity and at the same time for their stability and broad dynamic range, qualities that engineers often find to be anti-

* Originally published in Başar E (ed) Dynamics of sensory and cognitive processing by the brain. Springer, Berlin Heidelberg New York, pp 19–29 (Springer series in brain dynamics, vol 1). Cross references refer to that volume.

thetical. Olfactory sensitivity lies in part in the regenerative molecular feedback mechanisms of single cells (Lancet et al. 1985) such that a single odorant molecule may trigger a train of action potentials in a receptor cell. However, sensitivity is also provided, especially in macrosmatic animals, by the immense numbers of receptors. In the cat, for example, there are in the order of 10^8 receptors on each side, a numerosity enhances the likelihood of capture of molecules in turbulent air passed over the turbinate bones. Herein lies a major difficulty in understanding olfaction, which Lashley (1950) identified in vision as the problem of stimulus equivalence. Supposing that there might be in the order of 10–100 types of receptor, then there must be 1–10 million of each type. If an odorant can be identified repeatedly at concentrations ranging over 3–5 orders of magnitude, and if the lowest concentration involves stimulation of 10–100 receptors, how is an invariant constructed by the brain for an odorant over multiple trials, when the spatial pattern of excited receptors is never twice identical? The same type of problem obviously occurs in visual recognition of faces or signatures and auditory recognition of voices or words.

Sensitivity in olfaction is enhanced by experience under reinforcement and is disenhanced without it. Most of us have a limited repertoire of about 16 odorants under absolute discrimination, but the number can be increased without limit by sustained practice (Cain and Engen 1977). We can recognize some odors that were once important to us at intervals over many years in a flash flood of vivid associative memories that impel us to action. These are basic properties that olfaction shares with all other senses, far transcending in importance the decomposition of stimuli into lines, planes, and spectral peaks.

3 Neural Dynamics: Nonlinearity

I propose here that all of these properties inhere in the bulb in a single, comprehensive, nonlinear operation. The main bulbar constituents are large numbers of densely interconnected excitatory neurons (the mitral and tufted cells) and inhibitory neurons (the granule cells). The receptor input (Fig. 1) is spatially coarse-grained into segments corresponding to glomeruli, which form the bulbar equivalent of cortical columns with a mean segment width of about 0.25 mm. There are about 2000 glomeruli in each bulb of the rabbit. The several types of periglomerular interneurons in the outer layers of the bulb perform various janitorial tasks of input dynamic range compression, automatic volume control, spatial contrast enhancement, clipping, holding, and dc offset or bias regulation, among others (Freeman 1975). The negative feedback relation between the mitral and granule cells (Rall and Shepherd 1968) establishes a neural oscillator that receives its input through each glomerulus. These oscillators are coupled by mutually excitatory axosomatic synapses broadly over the bulb (Nicoll 1971) and by mutually inhibitory interactions through cellular mechanisms not yet clearly identified. Their output under coupling is at a frequency in the gamma range of 35–90 Hz, determined in the main by the passive membrane time constants (about 5 ms) and by the gains in the three types of feedback loop. Because of the widespread coupling, the EEGs from all parts of the

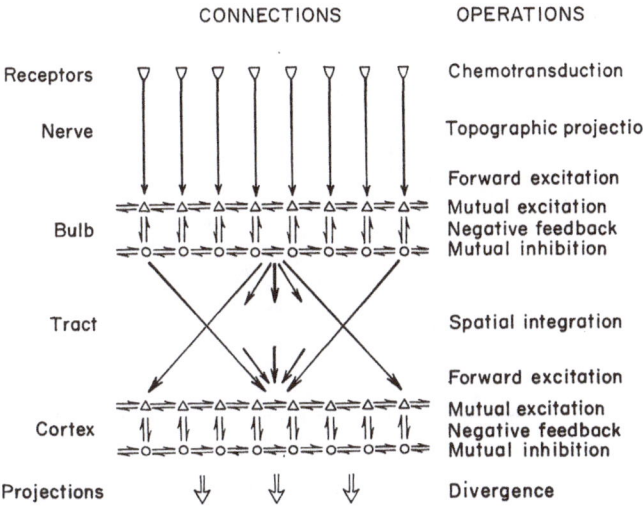

Fig. 1. A flowchart of activity in the olfactory system. Each layer is organized into a sheet of neurons. The state variables are defined for the axonal and dendritic modes in the two surface dimensions. They are discretized at intervals corresponding to the spatial coarse-graining by the glomeruli. Interactions occur laterally in each layer. The primary olfactory nerve provides for topographic projection of the input, whereas the lateral olfactory tract provides for spatial integration of the output. (From Freeman 1983)

main bulb at all times have a common waveform and everywhere a common instantaneous frequency (Freeman 1986).

These oscillators are inherently nonlinear. The nonlinearity stems from the voltage-dependent nonlinearity modeled for the action potential of nerve membrane by the Hodgkin-Huxley equations (Freeman 1979a). In the neural ensemble, it emerges as a sigmoidal function (Fig. 2) that relates pulse density (pulses per second per unit volume of the ensemble) to the density of excitatory dendritic current at the trigger zones. The curve is asymptotic to zero pulse density with inhibitory postsynaptic potential (IPSP) current and to a maximum for the ensemble with excitatory postsynaptic potential (EPSP) current. Two processes combine to give this shape. One is the exponential increase in tendency to fire with increasing depolarization (the sodium permeability or m-factor in the Hodgkin-Huxley equations). The second is the collection of metabolic, restorative, accommodative, and hyperpolarizing processes that establish the upper limit on firing rate, both on the long-term firing of single neurons and, by the ergodic hypothesis, on the entire ensemble over the short term. The nonlinearity is static, as distinct from the time-varying linear relationship that holds between membrane potential and firing rate for regularly firing single neurons. This is because neurons spend 99% of their lifespan below threshold, and because the firing pattern of each neuron closely resembles a Poisson process unrelated to those of its neighbors.

The nonlinear function is determined experimentally by calculating the pulse probability of mitral cell firing conditional on the EEG amplitude. The calculation is repeated for each EEG sample at 1 ms digitized intervals forward and backward in

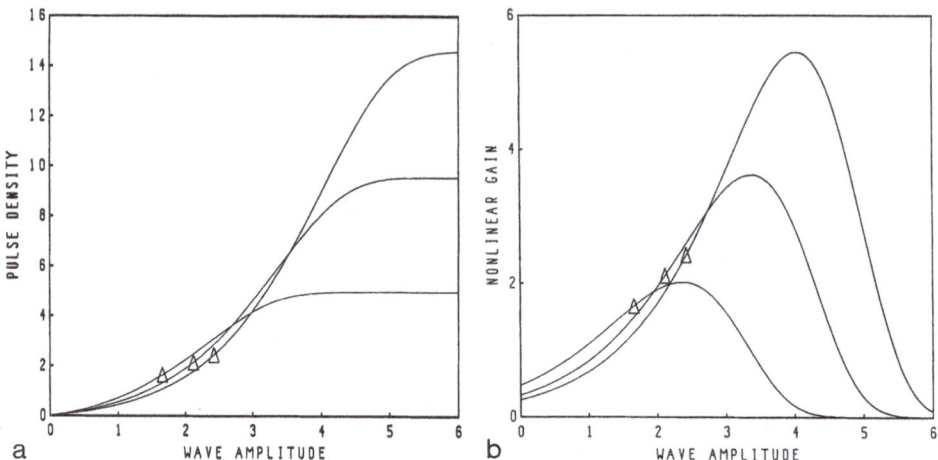

Fig. 2. (a) Three examples of a curve fitted to statistical data showing conversion of dendritic current density to axonal pulse density. (b) Derivatives of the three curves that give the nonlinear gain. *Triangles,* resting or equilibrium values. With increasing current amplitude there is a coupled increase in pulse density and in gain. (From Freeman 1979a)

time ± 25 ms, in order to allow for the time lags in the neural oscillator. The procedure also serves to demonstrate that the firing probability of each mitral cell oscillates at the common EEG frequency, and that the modulation amplitude in firing rate co-varies with the peak-to-peak amplitude of EEG oscillation. Mitral cell firing is statistically closely related to the EEG at all times and at each point of the bulb.

The nonlinear function for each bulbar ensemble is under centrifugal control. The shape of the sigmoid curve is retained, but the steepness is subject to increase, along with an increase both in mean and maximal firing rates. The derivative of the function represents the nonlinear gain of each local ensemble. The maximal gain is always displaced to the excitatory side. In animals under increased arousal or motivation, the gain is increased and the displacement to the excitatory side is extended, along with the increase in mean firing rate. The centrifugal input is most likely the cholinergic projection to the outer layers of the bulb. On the peripheral side, any receptor input excites the bulb and thereby raises its mean firing rate and its instantaneous gain. The curve is fixed but the operating point changes. Owing to the surge of receptor input with each inhalation, the bulb tends to undergo a recurrent increase and decrease in gain with the respiratory cycle.

Because of the bilateral saturation, the sigmoid curve is the most important mechanism providing for the stability of the bulbar mechanism (Freeman 1979b). The same curve also provides for its remarkable sensitivity, in the main because of the mutually excitatory feedback loop. Excitation of one subset excites another which re-excites the first, now in a more sensitive state, so that a regenerative increase in activity can occur. However, the negative feedback gain is also increased, so that instead of runaway excitation, a burst of oscillation appears. It begins during inhalation and ends during exhalation, and it is seen only in aroused, motivated animals (except occasionally in light stages of anesthesia, and then in an abnormal frequency range).

4 Neural Dynamics: Spatial Properties

Studies of the spatial patterns of these bursts manifested in the EEG have been made in rabbits with arrays of 64 electrodes chronically implanted over the lateral surface of the bulb. The EEG shows no dependence on novel odorants presented to naive animals, other than nonspecific changes associated with orienting responses. The spatial patterns of amplitude and phase modulation of the burst frequency vary within narrow limits about stereotypic mean patterns that are as characteristic for each individual as a handwritten signature. Under classical conditioning to respond differentially to two ordors (Viana di Prisco and Freeman 1985), one reinforced [conditioned stimulus (CS) +] and the other not (CS −), two new spatial patterns of amplitude emerge (Fig. 3), one for each CS. They are present only when the one or the other CS is present (Freeman 1986). For this demonstration, the EEG must be filtered with digital filters designed to conform to the spatial and temporal passbands

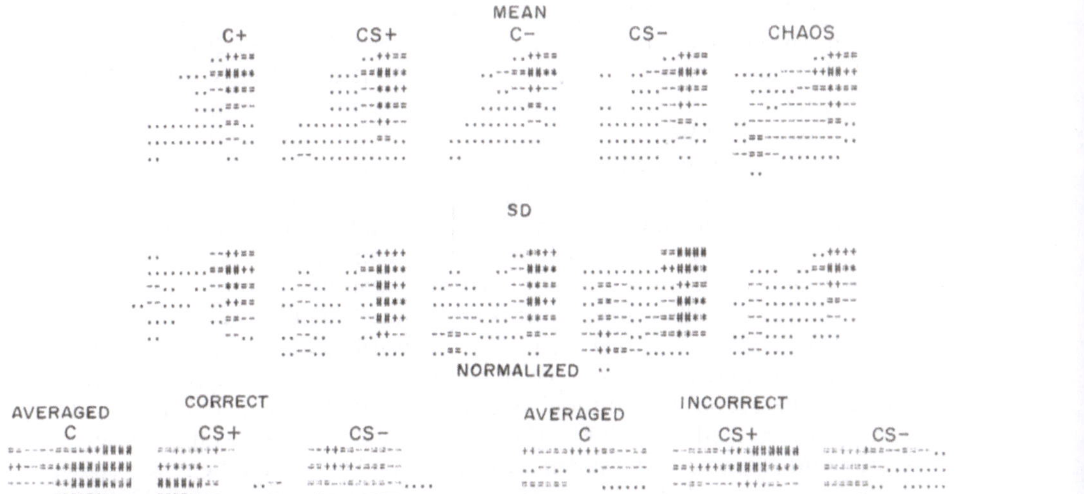

Fig. 3. Density plots (seven levels in descending order of amplitude #* + = − .) of EEG activity. *Upper frames*, means and SDs of amplitudes (*Chaos* refers to the disorderly bursts not subject to classification in respect to odors). *Lower frame*, amplitudes normalized by channel and by group, with those correctly classified on the *left* and those incorrectly classified by discriminant analysis on the *right*. *Bottom row*, patterns reconstructed from factor scores and loadings that were used for classification. (From Freeman 1986)

of the granule cell contribution to the EEG (Freeman 1986). The resultant patterns serve together with discriminant analysis to classify correctly, on average, 82% of EEG bursts sampled during control and test odor periods (Freeman 1986; Freeman and Viana di Prisco 1986). These patterns cover the entire array and, by inference from surface EEG phase gradients (Freeman 1986) and depth recording (Bressler 1984), the entire main bulb. The information density over the bulb is spatially uniform to within ± 5% (SD) of its mean, as measured by its value for correct classification of bursts.

The results show that insofar as the EEG is concerned, the bulb has the capability of responding selectively to odorants, but only in aroused animals that are trained to detect and respond to the test odors. This is in striking contrast to the results from unit studies in anesthetized or immobilized animals, which show selective responding of single neurons to some odors and not others, irrespective of training (e.g., Moulton 1976). Studies of metabolic activity with 2-deoxyglucose show that different patterns of radiographic density in the glomerular layer result from presentation of different odors (e.g., Lancet et al. 1982). These studies still lack proper controls for individual variation. The method allows only one odor for each animal; the EEG method shows foci of high amplitude activity that are similar to the high-density metabolic foci in size, shape, and location, but the degree of variation in EEG pattern between individuals exceeds that between odorants for each individual. Still, it is reasonable to conclude that input to the bulb from receptors establishes local regions of activity specific to an odor, and the output of the bulb is a global pattern involving all bulbar neurons, provided that the animal has been trained. Otherwise the global bulbar response is not spatially or temporally coherent or reproducible.

This transformation of local input to global output that incorporates past experience is the key to bulbar function. It is best understood by description in terms of nonlinear dynamics (Garfinkel 1983). A set of distributed, coupled, nonlinear oscillators has an infinite number of ways of performing, but within certain conditions of input and interaction strengths it tends to enter a definable state of activity and stay there until perturbed or modified. If under repeated perturbation it tends always to return to the same state, the system dynamics is said to have, or be governed by, an attractor. Attractors fall into three classes. The simplest is that of equilibrium; this occurs in the bulb only under deep anesthesia or in death. Periodic oscillation characterizes the limit cycle attractor; this appears in the EEG during bursts with inhalation. The most complex type is called the strange or chaotic attractor; its manifestation is nonperiodic activity that may appear to be random, of the sort that characterizes the resting EEG in nonmotivated animals and also the low-level EEG activity during exhalation.

Switching from one attractor to another is called a "state change" or "bifurcation". Its occurrence requires a parametric change in the system. Bulbar input provides for this by virtue of the nonlinear gain increase with receptor input during inhalation. The state change is from low-amplitude chaos to a high-amplitude spatially coherent limit cycle, and then back again. Order emerges from chaos and collapses with each cycle of respiration. There may be indefinitely many attractors of each type. Each is characterized by a set of parameter values and by a basin defined by a domain of input. The evidence suggests that a limit cycle attractor may form for each odorant that an animal is trained to respond to.

I believe that a limit cycle attractor is formed in the following way. On each inhalation of an odorant, the subset of the receptors that is sensitive to the odorant coactivates a subset of mitral cells. These are interconnected by excitatory axosomatic synapses that are bidirectional (Willey 1973). In accordance with Hebb's rule (Hebb 1949; Viana di Prisco 1984), these synapses are strengthened under coexcitation, provided that a reinforcing stimulus is paired with the odorant. Reinforcement activates neurons in the locus coeruleus, thus releasing into the bulb (and elsewhere) norepinephrine that enables the synaptic change (Gray et al. 1984). With repeated inhalations in the same and sequential trials, the odorant is delivered by turbulent flow in the nose to an ever-changing fraction of the subset of sensitive receptors, which leads progressively to the ultimate inclusion of all those mitral cells to which they project into a nerve cell assembly. These strengthened, mutually excitatory connections give the property to the assembly that, if any fraction of the sensitive receptors receives the odorant, their input to the bulb excites the entire assembly in a stereotypic manner (Freeman 1979c).

At once this constitutes figure completion, generalization over equivalent stimuli, and sensitization specific to a repeatedly reinforced class of stimulus. Computer simulations (Freeman 1979b) have shown that an increase of 40% on average in synaptic strength may increase the sensitivity of the bulb to a particular odorant by as much as 40000 times above the basal or naive level, because of the combination of mutual excitation and the nonlinear gain. After the completion of training, the subset of receptors activated during the training defines the basin of the attractor, and the nerve cell assembly of mitral cells determines the spatial structure of the limit cycle oscillation, which extends well beyond the assembly to involve the entire bulb. In principle, we can show how one odorant molecule can shape the activity of several hundred thousand second-order neurons.

I conceive the bulb as carrying a repertoire of learned limit cycle attractors, one for each odorant previously reinforced. Each is distinguished by its input basin with respect to receptors and by the spatial amplitude modulation pattern of its output. Random access is facilitated by the chaotic basal state, which keeps the bulb far from equilibrium and ready to move rapidly to any region of optimal convergence. The steadfast spatial pattern of bursts in the control state, in which no reinforced odor is given, indicates that an attractor exists for the background odor complex as well, and that bulbar output then signals the status quo. If a novel odor is given, the result is suppression of orderly burst activity and the appearance of broad-spectrum, spatially irregular, and nonreproducible bursts. Commonly, the highest peak of their multiply peaked spectra is at a frequency about half that of the sharply tuned frequency of the orderly bursts. I call these bursts "disorderly" or "chaotic". The prepyriform cortex to which the bulb projects responds to input as a tuned oscillator with spectral resonances around 18–24 Hz and 40–70 Hz (Freeman 1975). This suggests that the lower transmission frequency of the chaotic bursts can signal the failure of the bulbar mechanism to converge to a limit cycle attractor, and that repeated failures can lead to either of two outcomes: habituation if there is no reinforcement which updates sensitivity to a new status quo, or formation of a new limit cycle attractor under reinforcement. In other words, the bulbar mechanism provides a novelty detector without requiring an exhaustive search through information stored in the bulb.

5   Neural State Variables and Observables

Although the bulb has numerous specialized features not found elsewhere in the brain, these are not responsible for its main properties. At base it consists of a sheet of interconnected excitatory and inhibitory neurons with parallel input and output. This is an elementary description of neocortex as well. The static nonlinearity is a generalizable property of axonal membrane to be expected for every large ensemble in the cerebral cortex. The time and space constants are common to many, if not most, cerebral neurons. Hence the same basic dynamics can be expected to exist in all parts of the cerebral cortex.

I infer that odorant information is conveyed to the bulb by action potentials on particular receptor axons and that excitation is established and integrated among local subsets of mitral cells having apical dendrites within a limited number of glomeruli that correspond to neocortical columns. Following bifurcation, the entire bulb, comprising roughly $1 \, cm^2$ of cortical tissue, goes to a limit cycle attractor in the basin selected by the input. The output is global; the information is conveyed by action potentials on mitral axons, but it is in the form of a macroscopic pulse density function that is continuous in time and the two surface dimensions. The information is imposed as spatial amplitude modulation (in the surface dimensions as distinct from the time envelope) of the limit cycle carrier oscillation that is common to the entire bulb. Each event lasts in the order of 75–100 ms and repeats at the respiratory rate of 1–7 Hz. At the macroscopic level, each event can again be discretized into the surface grain of the glomeruli and the time frame of the burst; that is, olfaction can be treated as a sampled data system analogous to a digital graphic display.

The intrinsic state variables of a model for this system must correspond to the active states of pools of like neurons, which Sherrington identified as their central excitatory states (CES). For this reason, the proposed view might be described as neo-Sherringtonian. These activities are conveyed in local concentrations of action potentials, transmitter substances, and dendritic currents. They are manifested to observers in the forms of unit activity and electromagnetic field potentials. In all instances, the measurements of these observables must be properly filtered, averaged, and otherwise transformed in order to bring them into conformance with the CES, and they must be assigned to the proper elements in the model; for example, in the bulb, the EEG should be assigned to the granule cells and unit activity at the appropriate depth should be assigned to mitral cells.

The parallels to other sensory systems are straightforward. Information is conveyed by action potentials on thalamocortical axons and is established in local regions corresponding to columns, with different kinds of information being established at the microscopic level in each of the multiple cortical subareas comprising a sensory projection area. The neurons onto which the afferent activity is projected consist of excitatory and inhibitory neurons that are known to be densely interconnected by negative feedback and mutual excitation and can be inferred to have mutual inhibition as well. The crucial step for integration in perception may be the bifurcation of the interactive neural mass from a low-level chaotic attractor to a learned limit cycle attractor, such that the output of an extensive area of cortex at the macroscopic level might convey information on the whole in the spatial modulation of the amplitude of the limit cycle frequency. Again, input is local, output is global, and in analogy to the hologram, all parts of the output reflect all parts of the input.

Investigation of this hypothesis is likewise straightforward. The requisite carrier and gating frequencies respectively in the high beta and gamma ranges and in the theta and alpha ranges have been observed in most areas of neocortex. In visual cortex, during alpha suppression, the sequence of bifurcations requisite for trains of bursts might be provided by saccades. The steps that are needed to test the hypothesis are (1) the detailed spectral characterization of these activities, including use of complex demodulation over extended time series of the EEG; (2) the identification of the sources and sinks of the electric currents underlying these spectral peaks; (3) the assignment of these activities as states of variables of identified types of neurons in the cortex (4) measurement of the open loop time and space constants under deep anesthesia (Freeman 1975); (5) establishment of the spectral and spatial domains of neocortex over which commonality of wave form holds, such that chaotic or limit cycle attractors can be sought; and (6) behavioral analysis to determine the dimensions of the activity that relate to the stimulus and response variables selected for testing. Some progress has already been made in relating information content of visual and auditory stimuli to the waveforms of event-related potentials from neocortex. According to the present hypothesis, these correlations are adventitious and secondary, because the information relating to content is to be sought in the spatial dimensions, while the time courses of events are expected to reflect primarily the neural operations being performed on that information (Freeman and Schneider 1982).

None of these six steps is trivial; each may require several years to be brought to fruition. The outcome will be exceedingly important, because these kinds of information are essential to devise, evaluate, and improve macroscopic models of the distributed nonlinear dynamics of the forebrain.

In conclusion, the esence of cognition lies in forming and testing expectations based on past experience. In science it takes the form: if I do X, I expect A or B or the unexpected. Each outcome has predictable consequences. In rabbit olfaction it takes the form: if inhalation, then either status quo (the background), odor A, odor B, or an unexpected odor. Each inhalation is the action of a pattern generator or limit cycle attractor in the brain stem respiratory nuclei; each neural response is mediated by limit cycle attractors in the bulbs. I postulate that licking and sniffing are likewise mediated by limit cycle attractors in motor systems, whose basins receive the output of the bulbs. Basically this is a simple model of simple conditioned reflexes, but it tells us what to look for and how to look for it, as we try to understand how the brain synthesizes a percept from diverse sensory detail in the literal twinkling of an eye or the wriggle of a nose.

6 Summary

Neurons in cerebral cortex interact synaptically by mutual excitation, mutual inhibition, and negative feedback. Typically the negative feedback connections are locally dense, leading to the formation of local oscillators corresponding to columns. They are interconnected by mutually excitatory connections over large cortical areas. An appropriate model of cortex is a sheet of distributed coupled oscillators; observation is performed with arrays of surface EEG electrodes.

The dynamics of such systems are shaped by tendencies under perturbation to converge to stable states that are identified with attractors of three kinds. An *equilibrium* attractor is manifested in cortex by a steady state under deep anesthesia; a *limit-cycle* attractor is manifested by regular oscillation, and a *strange* attractor is manifested by chaos that appears to be random activity. Transition (bifurcation) from one attractor to another is imposed by a parametric change of the model or cortex.

The EEG of the olfactory bulb at rest appears chaotic. Inhalation excites the bulb, causing a parametric increase in negative feedback gain; the bulb bifurcates to a limit-cycle state. In the control condition of breathing air, the EEG spatial pattern is stereotypic for the background odor complex. With training to discriminate odors, a new spatial pattern appears with each odor, manifesting a learned limit-cycle attractor. These patterns appear to cover the entire bulb; input is local and output is global. The integration of a stimulus with past experience takes less than 0.1 s. Other sensory systems have similar properties; therefore bulbar dynamics may provide a useful model to explore preattentive processing in vision and other cognitive operations in the neocortex.

Acknowledgement. This work was supported by a grant MH06686 from the National Institute of Mental Health.

References

Bressler SL (1984) Spatial organization of EEGs from olfactory bulb and cortex. Electroencephalogr Clin Neurophysiol 57:270–276

Cain WS, Engen T (1977) Olfactory adaptation and the scaling of olfactory intensity. In: Pfaffman C (ed) Olfaction and taste III, Rockefeller University press, New York, pp 127–141

Efron R (1970) The minimum duration of a perception. Neuropsychologia 8:57–63

Freeman WJ (1975) Mass action in the nervous system. Academic, New York

Freeman WJ (1979a) Nonlinear gain mediating cortical stimulus-response relations. Biol Cybern 33:237–247

Freeman WJ (1979b) Nonlinear dynamics of paleocortex manifested in the olfactory EEG. Biol Cybern 35:21–37

Freeman WJ (1979c) EEG analysis gives model of neuronal template-matching mechanism for sensory search with olfactory bulb. Biol Cybern 35:221–234

Freeman WJ (1983) Dynamics of image formation by nerve cell assemblies. In: Başar E, Flohr H, Haken H, Mandell AJ (eds) Synergetics of the brain. Springer, Berlin Heidelberg New York, pp 102–121

Freeman WJ (1986) Analytic techniques used in the search for the physiological basis of the EEG. In: Gevins A, Remond A (eds) Methods of analysis of brain electrical and magnetic signals. Elsevier, Amsterdam (Handbook of encephalography and clinical neurophysiology, vol 3A/2)

Freeman WJ, Schneider WS (1982) Changes in spatial patterns of rabbit olfactory EEG with conditioning to odors. Psychophysiology 19:44–56

Freeman WJ, Viana di Prisco G (1986) EEG spatial pattern differences with discriminated odors manifest chaotic and limit cycle attractors in olfactory bulb of rabbits. Proceedings, conference on brain theory, Trieste 1984. Springer, Berlin Heidelberg New York Tokyo

Garfinkel A (1983) A mathematics for physiology. Am J Physiol 245:R455–R466

Gray CM, Freeman WJ, Skinner JE (1984) Associative changes in the spatial amplitude patterns of rabbit of olfactory EEG are norepinephrine dependent. Neurosci Abstr 10:121

Hebb DO (1949) The organization of behavior. Wiley, New York

Herrick CJ (1948) The brain of the tiger salamander. University of Chicago Press, Chicago

Lancet D, Greer CA, Kauer JS, Shepherd GM (1982) Mapping of odor-related neuronal activity in the olfactory bulb by high-resolution 2-deoxyglucose autoradiogrpahy. Proc Natl Acad Sci USA 79:670–674

Lancet D, Heldman J, Chen Z, Pace U (1985) Odorant-sensitive adenylate cyclase in olfactory cilia. Am Chem Soc Abstr 7

Lashley KS (1950) In search of the engram. Symp Soc Exp Biol 4:454–482

Moulton DG (1976) Spatial patterning of response to odors in the peripheral olfactory system. Physiol Rev 56:578–593

Nicoll RA (1971) Recurrent excitation of secondary olfactory neurons: a possible mechanism for signal amplification. Science 171:824–825

Rall W, Shepherd GM (1968) Theoretical reconstruction of field potentials and dendrodendritic synaptic interactions in olfactory bulb. J Neurophysiol 31:884–915

Viana di Prisco G (1984) Hebb synaptic plasticity. Prog Neurobiol 22:89–102

Viana di Prisco G, Freeman WJ (1985) Odor-related bulbar EEG spatial pattern analysis during appetitive conditioning in rabbits. Behav Neurosci 99:964–978

Willey TJ (1973) The ultrastructure of the cat olfactory bulb. J Comp Neurol 152:211–232

Self-Similarity in Hyperchaotic Data*

O. E. RÖSSLER and J. L. HUDSON

1 Introduction

The brain is a giant dynamic system. Its complexity should exceed e.g. that of a dripping faucet (Rössler 1977), which is one of the work-horses of nonlinear science (Martien and Shaw 1985). There are numerical methods for estimating the embedding dimension of a signal generated by a dynamic system (Froehling et al. 1981) which have been successfully applied to many kinds of data including brain-generated ones (see Mayer-Kress 1986 for many references).

From partial differential equations – which are nothing other than infinitely many coupled ordinary dynmic systems, all equal – it is known that frequently the dynamics involves but a few (finitely many) variables according to the center manifold theorem (Ruelle and Takens 1971). (Compare in this context a quiet lake on a wind-still evening: it is virtually at rest, that is, it has a point attractor.) Therefore it is conceivable in principle that the brain, with all of its 10^{11} neurons, 10^{14} synapses, and thousands of miles of cable (excitable PDEs; Winfree 1980), might likewise generate a cooperative signal that involves but a few variables (that is, dimensions) during part of its waking activity.

On the other hand, there is an important objection possible. A huge TV set – or a computer, for that matter – might also generate fairly low-dimensional (for example, multiple-periodic) signals when its internal circuitry is tapped either at a point chosen at random or in a global fashion analogous to scalp leads. These signals would clearly have to be interpreted as a "maintenance activity" that reveals almost nothing about the essential, meaningful activity of the machine in question. The latter activity would be determined by the machine's programs and/ or its visual inputs rather than by its internal clocks and scanning generators. The signals of the last-mentioned type might nevertheless be so preponderant, quantitatively, that the main activity only imposes itself as a sort of superimposed noise.

Similarly with the brain, low dimensional quasiperiodic signals – as discovered most clearly by Albano et al. (1987; see their beautifully resolved Fig. 5) – would have to be differentiated from a chaotic attractor of a similarly low dimensionality. Unfortunately, quasiperiodic signals contaminated by noise look a little like a chaotic attractor. To be more specific, the number of dimensions may be the same in both cases. Therefore, the question arises of how one might distinguish between these two extreme possibilities. As it turns out, the presently available methods of determining the dimension of a time series are not optimally

* Originally published in Başar E, Bullock TH (eds) Brain dynamics. Springer, Berlin Heidelberg New York, pp 113–121 (Springer series in brain dynamics, vol 2). Cross references refer to that volume.

suited to making this crucial distinction. There may be ways to improve their performance such that individual subdimensions governed by a periodic signal – possessing a close-to-zero underlying Lyapunov exponent – can be differentiated from other, genuinely chaotic subdimensions characterized by a positive Lyapunov exponent. This task, identification of the zero-LCE subdimensions, has, however, yet to be implemented.

An alternative possibility is to positively test for the presence of more or less large, positive Lyapunov characteristic exponents. Such positive exponents, while not proving chaos (Farmer 1981), nevertheless are a strong indicator that the dimension in question involves an unstable dynamics and thus may be part of a low-dimensional chaotic attractor. Again, the existing dimension-determining algorithms will have to be modified to permit picking up such subdimensions individually. While methods for identifying one or several positive exponents are available (Wolf et al. 1985; Stroop and Meier 1988), the just-proposed combination of breaking up the dimensions found into three subsets: trivial (periodic), nontrivial (chaotic), and a residue stemming from modulating (high-dimensional) noise, is apparently unavailable for the time being.

Interestingly, one of the three tasks of identification just described, the picking up of deterministic (hyper-)chaos contained in the data, can also be solved pictorially rather than algebraically. This new method is to be described in the following with an example.

2   The Self-Similar Fingerprints of Chaos

Chaos as is well known is closely related to the formation of a Cantor set. The latter constitutes the simplest self-similar fractal (Mandelbrot 1982). A chaotic attractor like that governed by the Hénon map (Hénon 1976),

$$x_{n+1} = a x_n (1 - x_n) - b y_n \tag{1}$$
$$y_{n+1} = x_n,$$

possesses a Cantor-like vertical cross-section (in the y direction) as is well known. Under linear magnification, as shown by Hénon in his original paper (1976), layers upon layers appear. However, this sandwich structure is not suitable for diagnostic purposes when the amount of noise is not negligible. The reason is that unless the attractor is very weakly attracting (that is, b, the constant contraction factor, is close to unity), the sheet-like structure is locally very thin ("1-D like"). Very little noise then suffices to blur the distinction between layers.

There is, however, a very robust "second" Cantor set hidden in the attractor, in the x direction. This fact was exploited by Smale (1967) when he first described his "basic set" (that is, the uncountable set of periodic solutions of differing periodicities hidden in such horseshoe-shaped maps). As shown in Fig. 1, the basin to the left of the saddle S "invades" the region to its right in a set of stripes that indeed "eat out" a Cantor set of vertical excisions. This happens in Eq. (1) whenever the parameter a exceeds a value close to 4, as is well known (Mira 1987). The attractor then of course also disappears (to let only the basic set survive). However, the very same structure even "lives" within the unexploded attractor.

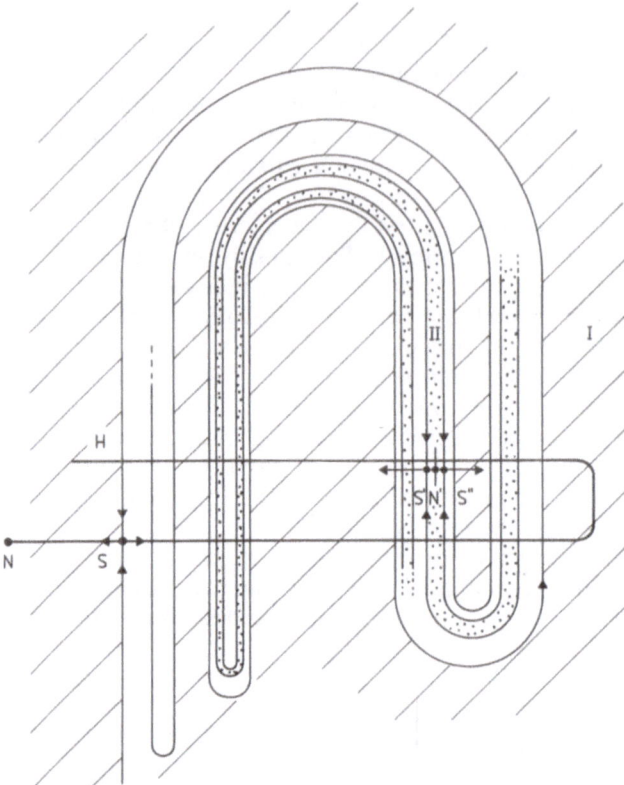

Fig. 1. Vertical "biting out of stripes" from an exploded former chaotic attractor. The attractor is exploded because (as happens in Eq. 1 with $a > 4$) the right-hand unstable manifold of the main saddle S has crossed the separatrix to the left-hand basin, at point H. The black (*dashed*) basin of the atractor N then "invades" the region of the former chaotic attractor, leaving only the "basic" set of periodic points (not shown) intact. Also shown is a second basin, of the attractor N', which is embedded in the former chaotic attractor. Both basins then intermingle in a very complicated fashion. (See Mira 1987, pp 296–297)

It can easily be recovered pictorially. This fact is shown in Fig. 2, (b), where we see the original Hénon attractor (a), rendered in an unusual way. Every point that next arrives in bin I has been plotted in black and every point that instead arrives in bin II first is shown in white (blank). Evidently, the "same" Cantor stripes as seen previously in Fig. 1 appear. The reason for this is that the dynamics of Eq. (1), with $a < 4$, still contains the Cantor structure of Fig. 1 "symbolically." The technique used in Fig. 2b is an example of symbolic dynamics. Symbolic dynamics looks at sequences of digits appearing in chaotic maps (see Crutchfield and Packard 1982). In the present case, it was instead used as a "coloring scheme."

Why is this possibility of staining data worth mentioning? The reason has to do with the fundamental stretching and folding process that underlies chaos (Rössler 1976; see Rössler 1983 and 1987a for pictures of a taffy puller). This repetitive stretching is responsible for the fact that, going back in time, every point

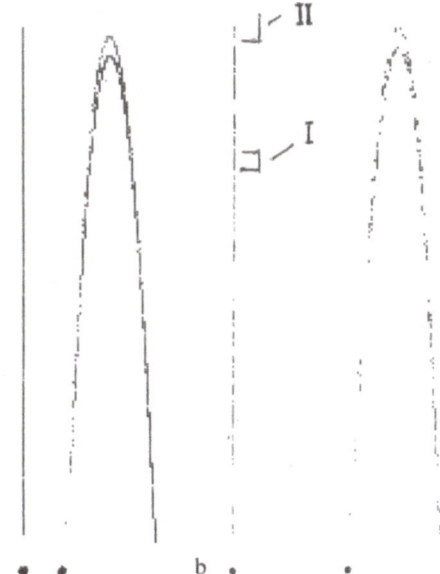

Fig. 2. (a) The Hénon attractor, Eq. (1) with $a=3.94$ and b positive ($b=0.05$). The x axis is shown here pointing up, whereas in Fig. 1 it pointed to the right. In this way, both a complete 2-D view to be compared with Fig. 1 can be presented (*on the right*) and a simplified but more informative 1-D top-down view (*on the left*), jointly as a single stereo pair. (The same convention will apply in Fig. 3.) The origin, point 0,0, is shown *bold*. Initial point 0.1, 0; 1200 iterates are shown at 11-digit accuracy. (b) The same data, replotted using a two-color convention. All points that the next time first land in bin I rather than bin II are shown in *black*, while those that do the opposite are erased (shown in white as *blanks*). Compare with Fig. 1. Bin I, $x>0.95$; bin II, $0.72<x<0.75$; 2300 iterates

in the "dough" can be classified according to how it will, in the future, behave relative to its neighbors. Hence any pair of "escape hatches" (regions or bins in the dough from which there is no return) necessarily have antecedents, characterizable by the "color" of the bin of their destiny, at every level of resolution, within the taffy. This fact is a direct consequence of the exponential mixing process (Rössler 1987 a) that takes place within the taffy (attractor). Unexpectedly, this result is robust when more than one dimension of stretching is involved. In this "hyperchaos"-generating situation (Rössler 1979), the pattern becomes even more distinctive. The simplest equation for an invertible map generating hyperchaos is a straightforward generalization of Hénon's 2-D map to three dimensions. It reads (Rössler et al. 1988):

$$x_{n+1}=ax_n(1-x_n)-by_n$$
$$y_{n+1}=ay_n(1-y_n)-bz_n \tag{2}$$
$$z_{n+1}=x_n.$$

This equation, too, is a globally smooth map (a diffeomorphism), whose constant contraction factor per step (Jacobian determinant) this time is equal to b^2 (rather than b). Its points could again have been generated by a smooth dynamic system, like the even simpler continuous hyperchaos generator of Rössler (1979) or its chemical analogue (Killory et al. 1987) for example.

Figure 3 b shows a picture analogous to that of Fig. 2 b. It was obtained using the two bins labelled I and II. Instead of a Cantor *stripes* structure, we have a Cantor *crosses* structure ("iterated Swiss flag"). If in place of the two bins chosen in Fig. 3 b, two other bins are chosen arbitrarily, the point patterns of Fig. 4 arise e.g. Thus, the fingerprint of hyperchaos is not a *particular* self-similar "unmixing" structure, but rather the fact that a self-similar structure arises at all whenever each point is forced to eventually go into one of two (or more) final patches.

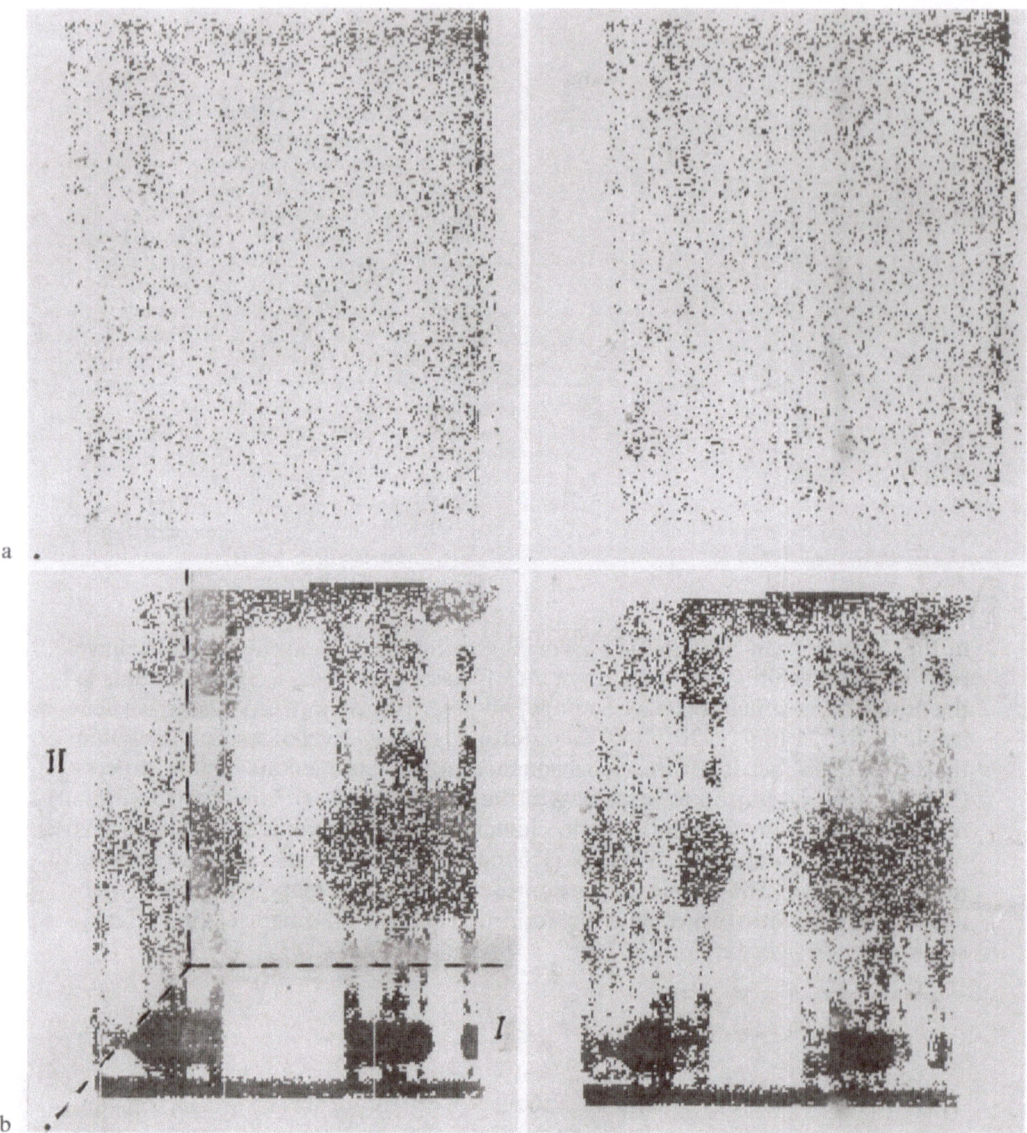

Fig. 3. (a) The hyperchaotic attractor, Eq. (2) with parameters a and b as in Fig. 2. Stereo pair as in Fig. 2. The *left-hand picture* is the top-down view (x, y), the *right-hand picture* has $-0.1\, z$ added to the horizontal component (x). Initial condition 0.1, 0, 0.1. The origin 0, 0, 0 is bold; 10 000 iterates. Eventually everything becomes black. (b) The same data, replotted using a two-color convention. As in Fig. 2 b dependent on whether the present string of points has bin I or bin II coming ahead next, it is made visible or suppressed, respectively. The prototype ("iterated Swiss flag"; see Rössler et al. 1986, 1988), obtaining for $b \to 0$, is still recognizable. Resolution can be improved by taking only a thin slice rather than the full z thickness. Bin I, $x < 0.3$ with $y < x$; bin II, $y < 0.3$ with $x < y$; 600 000 iterates. Note: Choosing the two parameters a differently in Eq. (2) will leave the result qualitatively unchanged

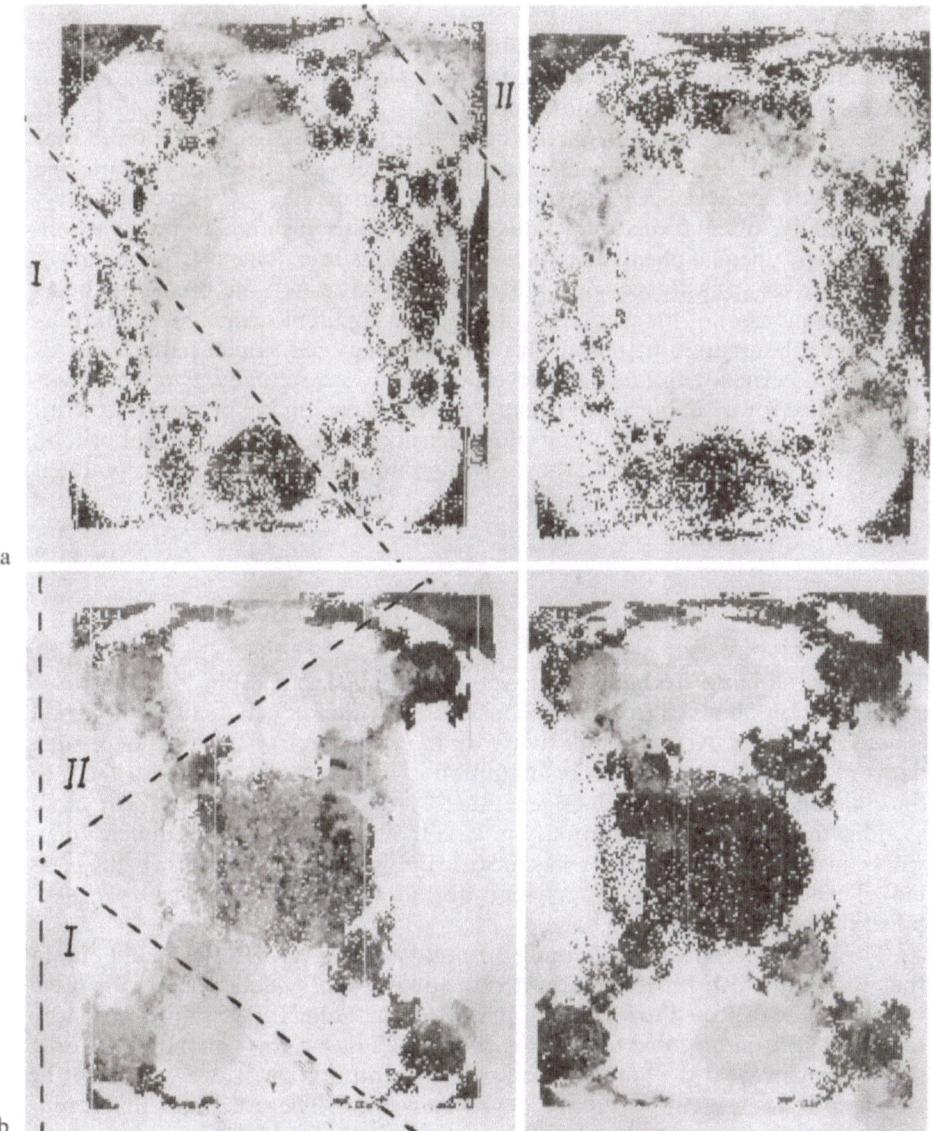

Fig. 4. (a) The data of Fig. 3a replotted in a different way. Bin I, $x+y<0.8$; bin II, $x+y>1.7$; 110 000 iterates. (b) The data of Fig. 3a replotted in a third way. Bin I, $y<0.5–0.6x$; bin II, $y>0.5+0.6x$; 320 000 iterates

3   Future Developments

The above examples are not yet real-life. Instead of maps, continuous flows will have to be used next to check how readily the self-similar structure survives in a Poincaré cross section when (a) the two Lyapunov characteristic exponents are no longer more or less equal (as was the case in Fig. 3), and (b) masking noise is added. Such data could be generated, for example by using the 4-variable system of Rössler (1979) calculated at low accuracy.

Nevertheless we expect the method to be fairly robust. Test runs showed that when the second sheet-like dimension contained a zero exponent, for example, the self-similarity indeed once more becomes locally stripe-like. Moreover, by varying the relative orientations and widths of the two bins, it should be possible even with strongly unequal positive Lyapunov characteristic exponents (stretching factors) to recover an easily recognizable self-similar pattern. If this prediction bears out, the main asset of the method could become apparent, namely its high resistance to contamination by noise. This is because any self-similar pattern can already be discerned with fairly few levels of repetition. Actually, two would virtually suffice for a "living proof" that a deterministic mixing process is at work in the data. Moreover by plotting point densities over a certain neighborhood rather than individual points, arbitrarily high noise levels become acceptable if only the data sequence is long enough.

4 Discussion

The above "staining" techniques has yet to be applied to real-life data, that is, to a cross-section obtained from the reconstructed attractor as the same time signal is used may times over with a delay (work in preparation). However, once such data are stored in a computer, by definition it will be as easy to subject them to the staining decision as the data generated by Eq. (2), for example. Also "cross sections through the cross section," in case the attractor is higher dimensional, will be just as easily obtained as was the case in Fig. 3 (with only a slice being plotted). The technique therefore appears to form a viable option to the experimentalist; this at least is what we hope.

This technical side of the staining proposal is, however, not the main point in the present context. More important is the question of whether or not the *brain* can indeed be expected to yield data that reveal a "hidden" self-similar structure in phase space (as opposed to the time domain; cf. Başar, this volume). Here only the future can yield the answer. However, to venture a guess, we would not be surprised if no clearcut positive evidence showed up. The reason for this scepticism was already given in the Introduction. While it would not be surprising to find a "maintenance activity" of periodic or quasiperiodic type in the brain, the possibility of a low-dimensional deterministic chaotic process being implied in the data has yet to be demonstrated. Of course, a maintenance activity of a low-dimensional chaotic type is also a possibility that cannot be ruled out a priori; compare this idea with the data on the smell brain (Freeman, this volume). In addition, we know that chaos is among the behavioral capabilities of whole animals (Todt 1969) – with Eq. 3 of Rössler (1980) forming an appropriate model system, cf. Rössler (1981) – but also of a larger system formed, e.g. by two coupled formal brains (Rössler 1987b). Moreover, it has been proposed that the brain as an organ may be subject to chaotic variations in its bulk activity (Braitenberg 1984), whereby, one might add, the local or not so local blood flow may form an essential participatory variable. Such "metabolic chaos" too could be considered as a type of chaotic maintenance activity of the organ in question. In a similar vein, it is conceivable that a chaotic pacemaker of hemodynamic and/or localized neuronal type (cf. Glass and Mackey 1979) may, in deep sleep or even in moments of idling, acquire the role of a "nonsignal" internally. Again, one could speak of a chaotic maintenance activity in that case. However, in spite of this possibility (cf. Rössler 1983), it would still come as a surprise if the fingerprints of low-dimensional hyperchaos of the type seen in Fig. 3 above could be shown to be present in scalp-EEG data, for example. In other words, the open question is whether or not the beautiful data available today (see Babloyantz, this volume) will be robust enough to allow for the detection beyond doubt of more than one positive Lyapunov characteristic exponent, either by using the present or by other techniques (e.g., Wolf et al. 1985; Stroop and Meier 1988).

To conclude, a new method has been proposed which may prove useful in the future analysis of real-life data generated by reaction-kinetic or biological systems that produce low-dimensional attractors. If everything goes well, the method may be of help in discriminating between deterministic hyperchaos generators, on the one hand, and other low-dimensional, but closer to quasiperiodic, time sequences of a similar phenomenological dimensionality on the other. The brain is but one out of many signal generators (cf. Bullock, this volume) for which one would like to have an answer to this type of question.

5 Summary

A new "staining technique" based on symbolic dynamics is proposed for investigating data generated by low-dimensional systems. The method is expected to help answer the question of whether or not the working brain is a generator of low-dimensional deterministic chaos.

Acknowledgments. We thank Walter Freeman, Erol Başar, Ted Bullock, Agnes Babloyantz, Paul Rapp, Gottfried Mayer-Kress, Jim Crutchfield, Helen Killory and Claus Kahlert for discussions. The work was supported in part by the N. S. F.

References

Albano AM, Mees AI, de Guzman GC, Rapp P (1987) Data requirements for reliable estimation of correlation dimensions. In: Degn H, Holden AV, Olsen LF (eds) Chaos in biological systems. Plenum, New York, pp 207–220

Braitenberg V (1984) Vehicles. MIT Press, Cambridge, p 65

Crutchfield JP, Packard NH (1982) Symbolic dynamics of one-dimensional maps: entropies, finite precision and noise. Int J Theor Phys 21:433–466

Farmer D (1981) Unpublished preprint

Froehling H, Crutchfield JP, Farmer D, Packard NH, Shaw R (1981) On determining the dimension of chaotic flows. Physica 3D:605–617

Glass L, Mackey MC (1979) Pathological conditions resulting from instabilities in physiological control systems. Ann NY Acad Sci 316:214–235

Hénon M (1976) A two-dimensional mapping with a strange attractor. Commun Math Phys 50:69–77

Killory H, Rossler OE, Hudson JL (1987) Higher chaos in a four-variable chemical reaction model. Phys Lett 122A:341–345

Mandelbrot B (1982) The fractal geometry of nature. Freeman, San Francisco

Martien P, Shaw R (1985) The dripping faucet. Phys Lett 110A:399–402

Mayer-Kress G (ed) (1986) Dimensions and entropies in chaotic systems. Quantification of complex behavior. Springer, Berlin Heidelberg New York (Springer series in synergetics, vol 32)

Mira C (1987) Chaotic dynamics, from the one-dimensional endomorphism to the two-dimensional diffeomorphism. World Scientific, Singapore

Rössler OE (1976) Chaotic behavior in simple reaction systems. Z Naturforsch 31 a:259–264

Rössler OE (1977) Chemical turbulence – a synopsis. In: Haken H (ed) Synergetics. Springer, Berlin Heidelberg New York (Springer series in synergetics, vol 1)

Rössler OE (1979) An equation for hyperchaos. Phys Lett 71A:155–157

Rössler OE (1980) Chaos and turbulence. In: Haken H (ed) Dynamics of synergetic systems. Springer, Berlin Heidelberg New York, pp 147–153 (Springer series in synergetics)

Rössler OE (1981) An artificial cognitive-plus-motivational system. Prog Theor Biol 6:147–160

Rössler OE (1983) The chaotic hierarchy. Z Naturforsch 38a:788–801

Rössler OE (1987a) Anaxagoras' idea of the infinitely exact chaos. In: Marx G (ed) Teaching nonlinear phenomena, vol 2. Chaos in education. Hungarian Publications in Physics Education, National Center for Educational Technology, Veszprém, Hungary, pp 99–113

Rössler OE (1987b) Chaos in coupled optimizers. Ann NY Acad Sci 504:229–240

Rössler OE, Kahlert C, Parisi J, Peinke J, Röhricht B (1986) Hyperchaos and Julia sets. Z Naturforsch 41a:819–822

Rössler OE, Klein M, Hudson JL, Wais R (1988) Self-similar basin boundary in an invertible system (folded-towel map). In: Kelso JAS, Mandell AJ, Shlesinger MF (eds) Dynamic patterns in complex systems. World Scientific, Singapore, pp 209–218

Ruelle D, Takens F (1971) On the nature of turbulence. Commun Math Phys 20:167–195

Smale S (1967) On differentiable dynamical systems. Bull Am Math Soc 73:747–829

Stroop R, Meier PF (1988) Evaluation of Lyapunov exponents and scaling function from time series. J Opt Soc Am[B]5:1037–1045

Todt DJ (1969) On the control of irregular behavior sequences: results of an analysis of the song of the blackbird, *Turdus merula* (in German) In: Marco H, Farber G (eds) Kybernetik 1968. Oldenbourg, Munich, pp 465–485

Winfree AT (1980) The geometry of biological time. Springer, Berlin Heidelberg New York

Wolf A, Swift JB, Swinney HL, Vastano JA (1985) Determining Lyapunov exponents from a time series. Physica 16D:285–317

Estimation of Correlation Dimensions from Single and Multichannel Recordings – A Critical View*

A. Babloyantz

1 Introduction

One of the most intriguing features of the nonlinear dynamics of the past decades is the realization that even deterministic dynamics involving only a few degrees of freedom may generate random behavior. The work of Packard et al. (1980) and Takens (1981) showed that from a seemingly random behavior, valuable information could be extracted for the characterization of the dynamics underlying the system which produces the signal. Algorithms were developed whereby from the measurement in time of a single parameter of a system, phase spaces could be constructed, and the correlation dimensions of the attractors, the Lyapunov exponents and the Kolmogorov entropies evaluated (Grassberger and Procaccia 1983a; Farmer et al. 1983, Wolf et al. 1985). Thus, one can distinguish between determnistic dynamics and random processes and classify systems according to their degree of coherence.

Such an approach opened a wide new area of investigation and at last one could tackle problems that were difficult to handle before. The investigations extended to such diverse fields as hydrodynamics (Brandstater et al. 1983), physics (Jeffries 1985), chemistry (Roux et al. 1983), climatic variability (Nicolis and Nicolis 1984), and physiology (Babloyantz et al. 1985).

Electroencephalograms (EEGs) in general show rather random behavior, although a given pattern may characterize a well-defined stage of cerebral activity. Therefore, it was natural to use these new formalisms of nonlinear dynamics for the elucidation of activities underlying the EEG.

During the first meeting of "Dynamics of Sensory and Cognitive Processing of the Brain," Babloyantz et al. (1985), Babloyantz (1988), and Başar et al. (1988) using the formalism of nonlinear dynamics showed that some stages of human and animal EEG may be described by deterministic chaotic dynamics involving only a few degrees of freedom and exhibiting a low value of correlation dimension.

Two years later, at a meeting of the same group, not only the number of papers dealing with deterministic chaos had increased substantially, but a workshop was devoted to the subject of nonlinear dynamics and its relevance to the study of the brain.

The aim of the present paper is to discuss the modality and validity of these methods of nonlinear dynamics for the study of the EEG. Among these methods spectral analysis, time autocorrelation function, and phase-space construction

---

* Originally published in Başar E, Bullock TH (eds) Brain dynamics. Springer, Berlin Heidelberg New York, pp 122–130 (Springer series in brain dynamics, vol 2). Cross references refer to that volume.

are qualitative methods, whereas Lyapunov exponents, Kolmogorov entropies, and the attractor dimension give quantitative information. Here we restrict ourselves to the evaluation of the correlation dimension using the by now standard algorithm proposed by Grassberger and Proccacia (1983 b). The concepts of fractal dimension, strange attractors, and Grassberger and Proccacia algorithms, could be found in Schuster (1984) and also in the paper by Başar et al. (this volume). We also show how one can extend the dimensional analysis to the multichannel recordings.

2 Phase-Spaces

The first step in any dynamic analysis is the construction of the phase portrait. Once the phase portrait is available, its topological dimension may be readily evaluated. In the absence of any knowledge about the details of the system's dynamics, a phase-space topologically equivalent to the actual system may be constructed from time series provided from the variation of a single parameter of the system.

The standard phase space construction necessitates only a single time series. All the relevant variables of the system are constructed from this time series. The construction relies on the fact that if $X(t)$ is the original time series, then the space spanned by the variables $X(t), X(t+\tau), X(t+2\tau) \ldots X(t+(n-1)\tau)$ is at least topologically equivalent to the original phase space and therefore represents many of its dynamic properties. Moreover, n is restricted by the relation $n > 2d+1$ where d is the lowest integer greater than the dimension of the attractor (Grassberger and Porcaccia 1983 b). We refer to such a procedure as the "lagging method."

The lagging method is simple and straightforward. However, although in principle, for infinite time series every value of the lag τ is acceptable, in reality great care must be taken in the choice of τ. We shall come back to this question in the next section where we show the importance of the value of τ in the determination of the correlation dimension.

Recently Eckman and Ruelle (1985) have conjectured that the phase space spanned by several simultaneous measurements of an observable variable in various sites of the system $X_1(r_1,t), X_2(r_2,t) \ldots X_1(r_n,t)$ may also yield a phase portrait which is topologically equivalent to the portrait obtained from the variables X_1, $\ldots X_m$ of the system. Such a phase space may be constructed very easily from multichannel EEG recordings. In this construction the lag τ disappears and is replaced by another subjective quantity, namely the interelectrode distance which is bounded by a minimum length imposed by the physical reality.

A more recent technique for phase-space construction using a single time series was introduced by Broomhead and King (1986). It is based on singular value decomposition which is a noise-reducing procedure. In essence, in this technique one diagonalizes the covariance matrix constructed from the phase-space vectors of the previous constructions and one obtains orthogonal eigen vectors which may be used to reconstruct a phase portrait and evaluate its correlation dimension. Singular values decomposition may be applied also to multichannel recordings.

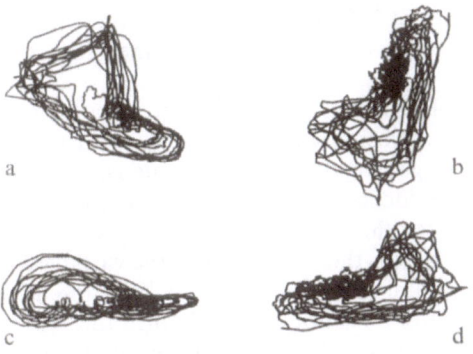

Fig. 1 a–d. Phase portraits of the Creutz-feld-Jakob attractor. A similar structure is seen from four different three-dimensional constructions: (a) Lagging method from one lead, (b) multichannel phase portrait, (c) and (d) are respectively the (a) and (b) portraits constructed using the singular eigen vectors

Figure 1 shows the phase portrait obtained from the EEG of a patient suffering from Creutzfeld-Jakob disease (Babloyantz and Destexhe 1988). All four procedures outlined above have been used. We note that the lagging method, multichannel portrait, multichannel data, after singular value decomposition, yield very similar phase portraits. On the contrary, the portrait obtained with the help of the singular eigen vectors and from a single time series (Fig. 1 c) looks different and is flat and smoother as noise and also probably some of the finer twists of chaotic trajectories have been eliminated.

3   Attractor Dimension

In order to discriminate between a deterministic and random activity, one evaluates the dimension D of the attractors which have been constructed in the phase space. Several definitions for the dimension exist. In this paper we shall concentrate on correlation dimension D_2 which can be evaluated for all attractors shown in Fig. 1.

The procedure for the evaluation of the correlation dimension from a single time series has been reported in several papers and will not be reproduced here. However, substantial error in the value of D_2 may appear if the recorded data have not been handled carefully. Several sources of error are possible.

1. Sampling frequency. The choice of the sampling frequency may be a source of error. Since the computation time approximately scales as $N \times (N-1)$, oversampling has to be avoided. In the case of spectral analysis, the usual rule adopted to avoid any loss of frequencies due to digitization is derived from the Shannon rule which says that the lowest sampling frequency is $2 \times f_{max}$, where f_{max} is the higher peak in the power spectrum. Can similar rules hold for dimensional analysis? We studied the same alpha waves sampled at three different frequencies (200, 400, and 1200 Hz) (Babloyantz and Destexhe 1986). In these cases, the power spectrum does not show significant peaks higher than 100 Hz. We observe that oversampling leads to substantial deformation of the spatial correlation integrals and to an underestimation of the correlation dimension. Recent algorithms have been proposed which prevent the onset of such deformations (Theiler 1986). We found that a sampling frequency of the order of 200 Hz and 15 s duration seems to be the most appropriate sampling rate for the evaluation of the dimension of the alpha waves.

2. *Time delay* τ. Although in principle all values of the time delay τ are acceptable for an infinite number of data points, in practice only a narrow range of τ will give correlation integrals with sufficiently large stretches of linear regions.

In spite of lack of rigorous theoretical evidence, procedures have been proposed in order to find an acceptable value for the delay τ (Fraser and Swinney 1986). These simply postulate that the best value is the one for which the variables are sufficiently uncorrelated such that minimum common information is present in the dynamics. Unfortunately this procedure does not always furnish the best τ especially for the EEG data. As the best value of τ, some authors prefer to choose the first zero of the autocorrelation function. In any case a check for stationarity of this parameter is strongly recommended. A wide range of values of τ must be considered before D_2 as a function of τ remains constant.

3. *Data length*. In order to get a good evaluation of D_2, ideally one must consider several hundred thousand points. However, this is not always possible, especially, in the case of the EEG where the dynamics do not remain stationary for very long. If the phenomenon is of short duration and if the dynamics are characterized by a rather low dimension, acceptable results may be found with several thousand points. In other cases, short time series may seriously underestimate the value of D_2. For example when analyzing the alpha waves according to the length of the time series, values as low as $D_2 = 2.6$ and as high as $D_2 = 6.6$ may be found. Here also one must use a data length such that a very substantial increase in the time series does not change the value of D_2. Of course the time must not exceed the duration of the stationarity of the phenomena.

4. *Upper limit of* D_2. When analyzing REM sleep and β waves which look like highly random signals, surprisingly, one finds respectively values of $D_2 = 9.7$ and $D_2 = 8.9$. We are not sure that such findings indicate the presence of a very highly deterministic chaos in these two stages of brain activity. Our suspicion is founded in the fact that even random white noise saturates at high embedding dimensions and gives a correlation dimension of the order of nine and ten. To be sure that the Grassberger and Proccacia algorithm does not saturate for random signals, calculations must be performed with time series completely free of any type of correlation. Until then we recommend that any value of $D_2 > 8$ be considered as suspicious.

If the abovecited precautions are taken into account, the correlation dimension may be evaluated reliably by digitizing the EEG. Such an analysis has been applied to the alpha waves, several stages of the sleep cycle, Creutzfeld-Jakob disease and petit mal type epilepsy. All these stages show the presence of deterministic chaos (Babloyantz 1988; Babloyantz et al. 1985; Babloyantz and Destexhe 1986; 1987 a, b, 1988 a).

The EEG probably manifests synchrony stages of neural masses. Low amplitude EEG reflects relatively desynchronized states, whereas high amplitude waves are an indication of synchrony between neural masses. In Fig. 2 the correlation dimension of these stages of the brain activity is plotted against a measure of the amplitude of the EEG. To obtain this measure, the signal is computed at 1-s in-

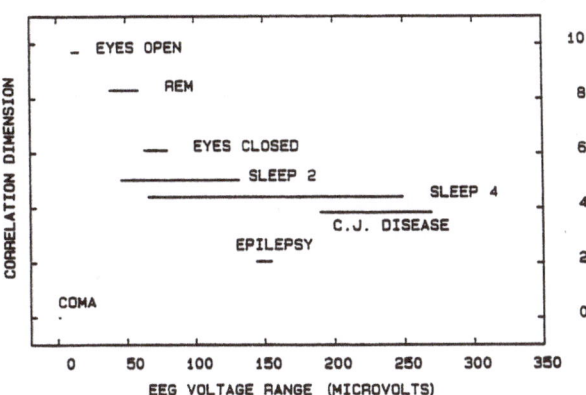

Fig. 2. Dimension-amplitude plot. The variations of EEG voltage observed at the optimal sampling frequencies (see text) and in successive stretches of 1 s are represented for each type of EEG behavior. The synchronization between neurons which occur in pathologies is reflected by high amplitudes and low dimensions

tervals. The highest and lowest amplitudes are included in the horizontal bars. During the normal states of brain activity, the dimension of the chaotic attractors decreases as the amplitude of the waves increases. The reverse situation is seen in pathological conditions. Creutzfeld-Jakob disease is characterized by a higher amplitude and a higher dimension than the epileptic seizure (Babloyantz and Destexhe 1987 a, b, 1988 a).

Multichannel recording is a common practice in most EEG laboratories. Therefore it would seem that multichannel analysis is much more economical and more suitable for study of the brain. However, here also one faces a new dilemma. How small or large should the interelectrode distance be?

If deterministic chaos is detected from a single-channel recording, it indicates the presence of chaotic activity in the recorded site. Nothing guarantees that the time series from an adjacent lead will show a chaotic activity or if it does, whether we are dealing with the previous chaotic attractor. Therefore, prior to the use of multichannel analysis, each channel must by studied by the lagging method to make sure that all of them belong to the same dynamics.

Presently we are engaged in such multichannel study of the EEG data. The analysis could be applied to multielectrode recording either in a horizontal setup or in a vertical insertion. In the case of the human EEG recorded from the scalp, many channels are contaminated with noise originating from muscular activity. This fact makes it difficult to find several channels that yield the same D_2 when the lagging technique is used. For example, in the case of Creutzfeld-Jakob disease, the values of D_2 in different electrodes range from $D_2 = 3.7$ to $D = 5.4$. This discrepancy may indicate that the disease does not involve a single attractor, or the unique attractor is highly contaminated with artifacts. If these leads are used in a multichannel analysis we find a value of $D_2 = 3.8$.

Better results could be obtained if one uses the EEG recorded in animal studies where electrodes are either inserted into the cortex or the activity is recorded from the surface of the cortex. However, in order to illustrate and test the procedure we consider multichannel electrical signals which are obtained by measuring the electrocardiogram (ECG). In this case there is no doubt that all leads must reflect the same dynamics, namely the activity of the cardiac muscle. The multichannel ECG of a normal individual, comprising 32 leads, is first analyzed chan-

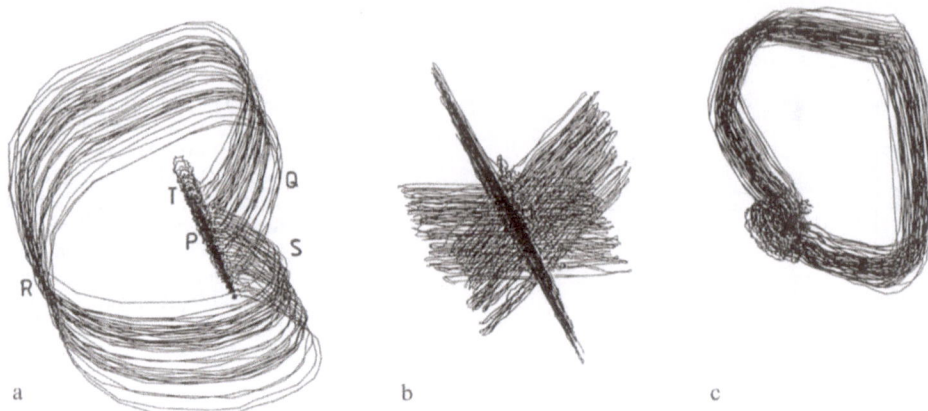

Fig. 3 a–c. Phase portraits of human ECG constructed in a three-dimensional space. A two-dimensional projection is displayed for two values of the delay, (a) 12 ms and (b) 120 ms; (c) represents the phase portrait constructed from the three simultaneous leads

nel by channel using the usual lagging method. Six leads of 240 s duration (60 000 points) gave $D_2 = 3.6 \pm 0.01$. Multichannel analysis was performed by using a combination of five leads chosen among those channels analysed by the lagging method. We found a value of $D_2 = 2.9 \pm 0.1$, which is in good agreement with previously reported results (Babloyantz and Destexhe 1988).

In this example and in all the other examples analyzed by us, we see a systematic discrepancy between the lagging technique and multichannel analysis. The latter always furnishes a value of D_2 smaller than the one obtained by the lagging method. The reason for this may be found in the following observation. The phase portrait obtained from a single channel using small τ has a striking resemblance to the one obtained from multichannel time series (see Fig. 3 a and c) whereas for the optimal τ the phase portrait is quite different (Fig. 3 b). The discrepancies between the two methods may be understood in the following manner. In the lagging procedure, as τ is increased, the attractor is gradually unfolded (the correlation intergrals show several simultaneous linear regions) showing its full complexity for the "best" τ after stationarity is reached. This gives a rather high value of D_2. On the contrary, when using space vectors, some of the variabilities of the dynamics are smoothed out and we observe a lower value of D_2.

The use of space vectors has many advantages over the usual lagging method: (a) one avoids costly trials before finding the appropriate time lag τ; (b) one sees a single, long and smooth linear region. Thus, the evaluation of the slope, D_2, is less ambiguous and more accurate.

4   A Visual Representation of Chaos

In the analysis of the physiological signals, sometimes, one may encounter almost periodic signals which, however, show obvious deviations from a purely limit cycle type of behavior. This is the case, for example, in the ECG. It is important

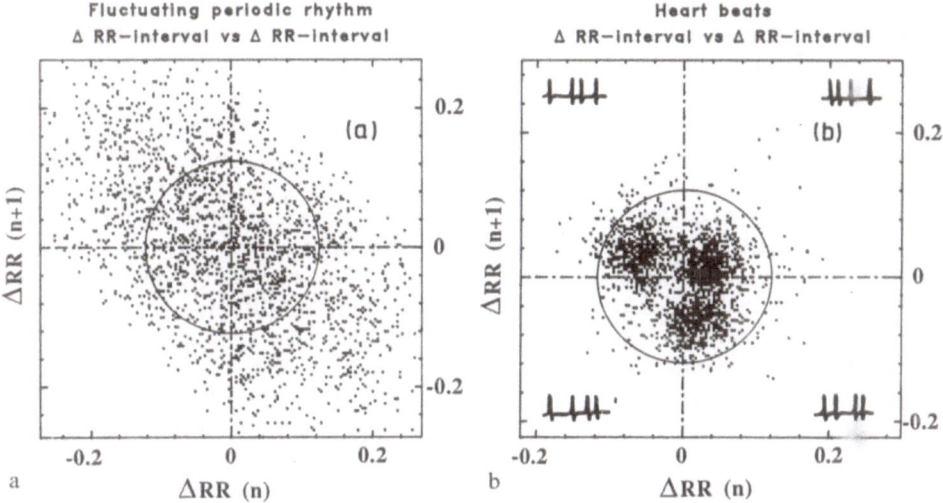

Fig. 4a, b. Representation of the interbeat interval correlation. The difference between two successive beat-to-beat intervals is plotted against the same difference one beat later. (a) Pseudo-random fluctuations around a periode oscillator. (b) Behavior of a normal heart. We see an obvious correlation between the variabilities of three consecutive RR intervals. (A digital filtering of the ECG below 0.5 Hz before the extraction of the RR intervals does not alter this structure)

to distinguish whether one is in the presence of a deterministic chaos or the purely periodic signal is contaminated by random noise. Such a distinction is possible with the help of the following analysis:

Let $RR(n)$ represent the time interval between two successive peaks of the almost periodic signal. One constructs a map of interval variation $\Delta RR(n+1)$ versus $\Delta RR(n)$. This construct reveals the correlations between the three successive peaks. For a perfect oscillator, the origin is the only possible value of ΔRR, whereas a noise-prone oscillator will exhibit a uniform distribution (Fig. 4) reflecting the statistical independence of the successive values of ΔRR. Figure 3b maps the variability of the normal heart. An obvious "structure" is seen in the fluctuations of the RR intervals which is an unambiguous sign of coherence in the interbeat variability (Babloyantz and Destexhe 1988).

5   Discussion

Brain activity is a time-evolving process and the only way to extract information from the EEG is to use methods which look at the dynamic aspect of the cerebral cortex. The nonlinear dynamics have furnished several methods of analysis suitable for such a study. In this paper we have only concentrated on the correlation dimensional analysis. However, today, from the same time series, we may compute other dynamic quantities such as autocorrelation functions, Kolmogorov entropies, Lyapunov exponents, and also construct Poincaré sections which are usable only for low-dimensional attractors. All these methods must be used together with dimensional analyses outlined in this paper. As we saw, the latter may give divergent results if various parameters of the time series are not chosen with care.

With this type of analysis, from the time measurement of a single variable, it is possible to assess the degree of coherence of the system, to discriminate between noise and deterministic dynamics, and to establish a hierarchy between several states of the same cortex. The finding that several key stages of brain activity follow deterministic chaotic dynamics sheds some light on the mechanism of information-processing capabilities of the cortex. This capability stems from the fact that chaotic attractors are sensitive to the initial conditions.

Moreover, these powerful tools of data analysis may help to bridge the gap between experimental and theoretical approaches to complex networks. It imposes constraints on model construction. A given model must finally reproduce dynamic properties of the real system as computed from the notions mentioned above.

The principal aims of this paper have been: (a) to show the pitfalls of the usual lagging method and how to avoid them; (b) to test for the first time, using physiological data, a conjecture put forward whereby one avoids some of the difficulties of the lagging method by using multichannel recordings; (c) to apply to physiological data the singular decomposition technique; and (d) to introduce a new way of characterizing more regular signals. However, research in this promising field of nonlinear dynamics is progressing rapidly. Algorithms exist which enable the evaluation of the number of local dimensions of attractors. A comparative study of all the existing methods and their use in the study of physiological signals may be found in a recent paper by Destexhe, Sepulcre, and Babloyantz (1988).

6 Summary

Recent developments in the theory of nonlinear dynamics provide several means for the study of complex systems. From the measurements of a single variable in the form of a time series, several important properties of the system can be assessed. Phase space construction, time autocorrelation function, and spectral analysis are qualitative methods, whereas correlation dimension, Lyapunov exponents and Kolmogorov entropy may be evaluated quantitatively. We discuss the relevance of such an approach to the study of EEG, evaluating the advantages and shortcomings of existing algorithms. We show the sensitivity of the correlation dimensions to the choice of the length of the data sets and also to the value of the lag. We applied these methods to the study of various stages of EEG in normal as well as in pathological instances. In normal individuals we show that as the wake-sleep cycle sets in, some stages of the brain activity oscillate between several deterministic dynamics of chaotic nature characterized by various degrees of coherence. The β waves need more than ten degrees of freedom for their description, whereas α waves show a chaotic attractor of rather high correlation dimension $D_2 = 6.1$ which may be described by deterministic dynamics with few degrees of freedom. Sleep stages two and four are respectively characterized by chaotic attractors with $D_2 = 5.03$ and $D_2 = 4.05$–4.40. As REM sleep sets in, again the coherence of the brain is lost and chaotic attractors vanish. In the case of severe pathologies such as coma and "petit mal" type epilepsy, the dimension of the chaotic attractor drops to a low value of $D_2 = 2.05$.

References

Babloyantz A (1988) Chaotic dynamics in the brain activity. In: Başar E (ed) Dynamics of sensory and cognitive processing of the brain. Springer, Berlin Heidelberg New York (Springer series in brain dynamics, vol 1)

Babloyantz'A, Destexhe A (1986) Low dimensional chaos in an instance of epilepsy. Proc Natl Acad Sci USA 83:3513

Babloyantz A, Destexhe A (1987a) Strange attractors in the human brain. In: Rensing L, an der Heiden U, Mackey MC (eds) Temporal disorder in human oscillatory systems. Springer, Berlin Heidelberg New York, p 48 (Springer series in synergetics, vol 36)

Babloyantz A, Destexhe A (1987b) Chaos in neural networks. In: Proceedings of the first IEEE international conference on neural networks, San Diego, June 1987

Babloyantz A, Destexhe A (1988a) The Creutzfeld-Jakob disease in the hierarchy of chaotic attractors. In: Markus M, Müller S, Nicolis G (eds) From chemical to biological organization. Springer, Berlin Heidelberg New York, p 307 vol 39

Babloyantz A, Destexhe A (1988b) Is the normal heart a periodic oscillator? Biol Cybern 58:203

Babloyantz A, Nicolis C, Salazar M (1985) Evidence of chaotic dynamics during the sleep cycle. Phys Lett 111 A:152

Brandstater A, Swift J, Swinney HL, Wolf A, Farmer JD, Jen E, Crutchfield JP (1983) Low dimensional chaos in a hydrodynamic system. Phys Rev Lett 51:1442

Broomhead DS, King G (1986) Extracting qualitative dynamics from experimental data. Physica 20D:217

Destexhe A, Sepulchre JH, Babloyantz A (1988) A comparative study of the experimental quantification of deterministic chaos. Phys Lett [A]132:101

Eckmann JP, Ruelle D (1985) Ergodic theory of chaos and strange attractors. Rev Mod Phys 57:617

Farmer JD, Ott E, Yorke JA (1983) The dimension of chaotic attractors. Physica 7D:153

Fraser AM, Swinney HL (1986) Using mutual information to find independent coordinates of strange attractors. Phys Rev 38A:1134

Freeman WJ (1986) Petit mal seizure spikes in a factory bulb and cortex caused by runaway inhibition after exhaustion by excitation. Brain Res Rev II:259

Grassberger P, Procaccia I (1983a) Characterization of strange attractors. Phys Rev Lett 50:346

Grassberger P, Porcaccia I (1983b) Measuring the strangeness of strange attractors. Physica 9D:189

Nicolis C, Nicolis G (1984) Is there a climatic attractor? Nature 311:529

Packard NH, Crutchfield JP, Farmer JD, Shaw RS (1980) Geometry from a time series. Phys Rev Lett 45:712

Roshke J, Başar E (1988) The EEG is not a simple noise: strange attractors in intracranial structures. In: Başar E (ed) Dynamics of sensory and cognitive processing by the brain. Springer, Berlin Heidelberg New York (Springer series in brain dynamics, vol 1)

Roux JC, Simoyi RM, Swinney HL (1983) Observation of a strange attraction. Physica 8D:257

Shuster H (1984) Deterministic chaos. Physik, Weinheim

Takens F (1981) Detecting strange attractors in turbulence. In: Rand DA, Young LS (eds) Dynamical systems and turbulence, Warwick 1980. Springer, Berlin Heidelberg New York, p 366 (Lectures notes in mathematics, vol 898)

Theiler J (1986) Spurious dimensions from correlation algorithms applied to limited time series data. Phys Rev 34A:2427

Wolf A, Swift JB, Swinney HL, Vastano JA (1985) Determining Lyapunov exponents from a time series. Physica 16D:285

Correlation Dimensions in Various Parts of Cat and Human Brain in Different States*

J. Röschke and E. Başar

1 Introduction

The analysis of deterministic chaos is currently an active field in many branches of research. Mathematically, all nonlinear dynamical systems with more than two degrees of freedom can generate chaos and, therefore, become unpredictable over a longer time scale. In order to describe periodic, aperiodic, or even chaotic behavior of nonlinear systems, several approaches have been used. In 1963 Lorenz applied concepts of nonlinear dynamics to the convection phenomenon in hydrodynamics in order to describe atmospheric turbulence (Navier-Stokes equation). He demonstrated the possibility that the unpredictable or chaotic behavior observed in an infinite-dimensional system might be caused by a three-dimensional dynamical system. Our research group's first nonlinear approach consisted in comparing the relation between EEG and evoked potentials by considering the Duffing equation as an adequate nonlinear model (Başar 1980). Later, assuming the EEG to be a chaotic attractor and mentioning the possibilities of applying the Navier-Stokes equation for comparison, we described that the EEG might reflect properties of a strange attractor (Başar 1983; Başar and Röschke 1983).

Since the pioneering analysis by Babloyantz et al. (1985), several investigators have undertaken the nonlinear analysis of the EEG using Grassberger and Procaccia's algorithm in order to evaluate the correlation dimension D_2 (see for example Babloyantz et al., Saermark et al., Graf and Elbert, Skinner et al., this volume). Our first analysis involved the evaluation of D_2 for EEG activities of intracranial structures of the cat brain (Röschke and Başar 1985 a, b, 1988; Başar et al. 1988). Although the present study also includes our earlier results in slow-wave sleep (SWS) for comparative analysis, our major aim was to look at a new frequency window – the range between 100 and 1000 Hz. In this frequency range the spontaneous activity of the cerebellum and the brain stem depicts ample electrical activity and resonances evoked by sensory stimulation (Adrian 1935; Başar 1980). Further, a first evaluation of D_2 during the transition stages was undertaken. An important trend in our analysis is also description of the dimensionality of EEG including the large fluctuations of D_2 during transitions. However, we submit that the use of other descriptors (for example, description by means of power spectra) is necessary to complement the interpretation of dimensionality.

* Originally published in Başar E, Bullock TH (eds) Brain dynamics. Springer, Berlin Heidelberg New York, pp 131–148 (Springer series in brain dynamics, vol 2). Cross references refer to that volume.

2   Definitions, Concepts and Explanations

A set of first-order differential equations

$$\dot{X}_i = F_i(x_i, t) \quad (i = 1, \ldots, n)$$

is called a dynamical system. If time does not appear explicitly in the function F_i, the system is called *autonomous*. It is a well-known fact that the nonlinearity of the F_i is a necessary (but not sufficient) condition for the generation of deterministic chaos. Deterministic chaos means that the behavior of the system is not predictable over longer periods. Nevertheless, there exists a prescription, for example, in terms of differential equations, for calculating future behavior from given initial conditions. Often investigations of dynamical systems have been performed in so called phase-space.

In general, phase-space is identified with a topological manifold. An n-dimensional phase-space is spanned by a set of n-independent linear vectors. This requirement is usually sufficient. Following a proposal made by Takens (1981) we span a 10-dimensional phase-space by x(t), ..., x($t+9\tau$) where τ is a fixed time increment. Every instantaneous state of a system is therefore represented by a set $(x_1, \ldots, x_n)$, which defines a point in the phase-space. The sequence of such states over the time scale defines a curve in the phase-space, called a trajectory. As time increases, the trajectories either penetrate the entire phase-space or they converge to a lower dimensional subset (Fig. 1). In this latter case, the set to which the trajectories converge is called an attractor.

In relation to the topological dimension of the remaining attractor, one can deduce various properties of the investigated system. If the Hausdorff dimension of an attractor is a noninteger, the attractor is called a strange attractor. All strange attractors which have been encountered up to now have a fractal dimension. In order to give a geometric explanation of the computation of the dimension, we will illustrate schematically a proposal made by Grassberger and Procaccia (1983) to compute the so-called correlation dimension D_2. Assuming that the

PHASE- SPACE

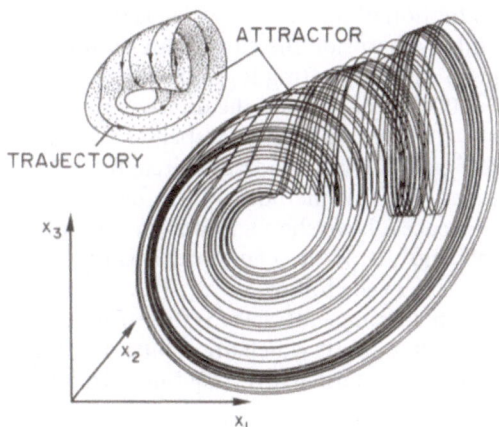

Fig. 1. Trajectories converge to a lower dimensional subset called an attractor. (Modified after Abraham and Shaw 1983)

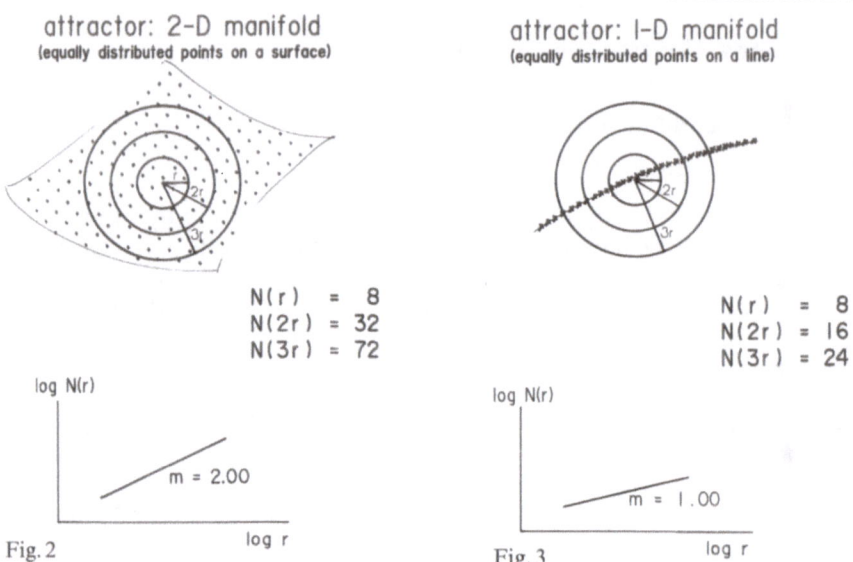

attractor: 2-D manifold
(equally distributed points on a surface)

attractor: I-D manifold
(equally distributed points on a line)

N(r) = 8
N(2r) = 32
N(3r) = 72

N(r) = 8
N(2r) = 16
N(3r) = 24

log N(r)

m = 2.00

log r

log N(r)

m = 1.00

log r

Fig. 2

Fig. 3

Fig. 2. For every point of the 2-D attractor the number N(r) of points lying inside a circle of radius r are counted. By plotting log N(r) versus log r a straight line with slope $m = 2.00$ is obtained

Fig. 3. By counting the number of points N(r) lying inside a circle of radius r and plotting log N(r) versus log r a straight line with slope $m = 1.00$ is obtained

attractor represents a two-dimensional manifold in the phase-space, it is possible to evaluate the dimension in the following manner. Figure 2 illustrates the basic features of the computational algorithm. For every point of the attractor, the number of points lying inside a circle are counted (or in the case of a three-dimensional phase-space inside a ball) which have a radius of r_0, $2r_0$, $3r_0$, etc. For the two-dimensional manifold $N(r_0) = 8$, $N(2r_0) = 32$, $N(3r_0) = 72$, ... Now, if a plot of log (r) versus log N(r) is performed, a straight line is registered. The slope of this line is $m = 2.00$. This is exactly the dimension of the attractor. This result never changes, even if a higher dimensional phase-space is considered.

It can be shown that if the attractor is not a two-dimensional manifold but a simple curve in the phase-space, the evaluation of the dimension leads to a value of $D_2 = 1.0$ (Fig. 3). The number of points lying inside the circle with radius r is $N(r_0) = 8$, $N(2r_0) = 16$, $N(3r_0) = 24$, and so on. By plotting log N(r) versus log r, one finds a straight line with slope $m = 1.00$. This is exactly the dimension of the simple curve, which we assumed to be the attractor. In fact, Grassberger and Procaccia's algorithm counts the number of points lying inside the circle for every point of the attractor and averages the results. This method is therefore simple, but time-consuming. Grassberger and Procaccia have shown that it gives reliable results, even though the attractor is a strange attractor. Therefore, it is possible to calculate fractal dimensions as we describe in the following.

One of the most important properties of a strange attractor is its sensitive dependence on initial conditions. This means that points which are arbitrarily close initially, become macroscopically separated after sufficiently long times. In other

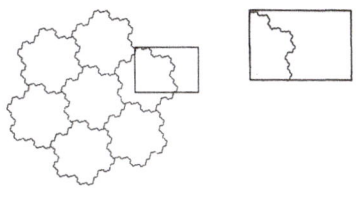

D = 1.129....

$7^{1/2} = 3^{1/1.129....}$

Fig. 4. A self-similar object with a fractal dimension of $D = 1.129$ ... (After Schroeder 1986)

words, similar causes do not produce similar effects. This is an extensive statement, which seems to damage the causality principle of natural philosophy. However, by examining the properties of a strange attractor more precisely, one finds that a strange attractor may have a strong conformity, called self-similarity, which is an invariance with respect to scaling.

Self-similar objects possess a fractal dimension. And "fractals," so called because their fractal or Hausdorff dimension exceeds their topological dimension and show strange properties, which are not known in linear systems (see Schuster 1984).

Let us now consider Fig. 4. The main pattern is obviously similar to its seven parts. The area ratio is $7:1$ and the perimeter is three fold the perimeter of one of the seven similar pieces (Schroeder 1986).

Euclid's geometry tells us that areas of similar figures are proportional to their linear dimensions squared. However, 3 squared equals 9 and not 7. Nowadays, we have to point out that geometric objects like the one shown in Fig. 4 do not obey Euclid's rules of geometry. The boundaries of the illustration in Fig. 4, while everywhere continous, are nowhere differentiable. This object is a fractal with a Hausdorff or fractal dimension of $D = 1.129$ ...

In this case we can reformulate Euclid's theorem: For similar figures, the ratios of corresponding measures are equal when reduced to the same dimension. In our case:

$$7^{1/2} = 3^{1/1.129}$$

Another example, which shows the strange properties of self-similar objects is presented in the following digital set:

10010110011010010...

This infinite binary sequence has the property that it is self-similar. In fact, striking out every second term the sequence reproduces itself

1001011001...

Surprisingly, although the binary sequence shown is aperiodic, its Fourier transform is not at all noise-like (Fig. 5). The self-similarity of the sequence induces

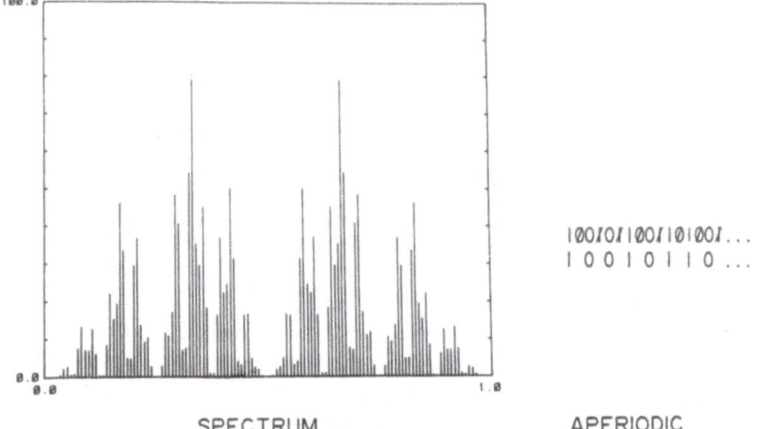

Fig. 5. Spectrum of the self-similar sequence (after Schroeder 1986), striking out every second term, the sequence reproduces itself

long-range correlations and causes its spectrum to be highly structured with many peaks as if it were periodic. In other words, self-similar sequences can be both aperiodic and have spectra resembling those of a periodic sequence.

3 Methods

3.1 Mathematical Methods

One of the oldest concepts of dimension is that of a topological dimension D_T. For a point, $D_T = 0$; for a line, $D_T = 1$; and for a plane $D_T = 2$. A first generalization is the Hausdorff dimension or fractal dimension D_F. For simple sets, for example, a limit cycle or a torus, the fractal dimension D_F is an integer and is equal to the topological dimension D_T. For an n-dimensional phase-space, let $N(\varepsilon)$ be the number of n-dimensional balls (or cubes) of radius ε required to cover an attractor. Then the fractal dimension D_F is defined as

$$D_F = \lim_{\varepsilon \to 0} \frac{\log N(\varepsilon)}{|\log \varepsilon|}$$

For the self-similar geometric object (see Fig. 4), one can choose: $\varepsilon = (\frac{1}{3})^n$, then $N(\varepsilon) = (1/\sqrt{7})^n$ and it follows that

$$D_T = \lim_{\varepsilon \to 0} \frac{\log N(\varepsilon)}{|\log \varepsilon|} = \lim_{n \to \infty} \frac{\log 3^n}{\log \sqrt{7^n}} = \frac{\log 3}{\log \sqrt{7}} = 1.12915\ldots$$

A generalization of the fractal dimension is introduced in information theory. The Renyi information of order q is defined as

$$I_q = \frac{1}{1-q} \log \sum_{i=1}^{N(\varepsilon)} p_i^q \quad (q \neq 1)$$

$$I_q = - \sum_{i=1}^{N(\varepsilon)} p_i \log p_i \quad (q = 1)$$

Let p_i be the probability that an arbitrary point (of an attractor) falls into ball i with radius ε and let $N(\varepsilon)$ be the number of nonempty balls. The generalized dimension D_q of order q is given by

$$D_q = \lim_{\varepsilon \to 0} \frac{I_q(\varepsilon)}{\log 1/\varepsilon}$$

For $q=0$ we find $D_0 = D_F$. D_1 is called the "information dimension" and D_2 is called the "correlation dimension." In practice, the correlation dimension D_2 is the generalized dimension easiest to be estimated from attractors generated by experimental data, because

$$I_2 = -\log \sum_{i=1}^{N(\varepsilon)} p_i^2 = -\log C(R)$$

where $C(R)$ is a measure of the probability that two arbitrary points x_i, x_j will be separated by distance R. $C(r)$ is called the "correlation integral" and can be easily computed:

$$C(R) = \lim_{N \to \infty} \frac{1}{N^2} \sum_{\substack{j,i=1 \\ \neq j}}^{N} \theta(R - |x_i - x_j|)$$

where θ is the Heavyside function.
It then follows that

$$D_2 = \lim_{R \to 0} \frac{\log C(R)}{\log R}$$

or $C(R) \propto R^{D_2}$.

The main point is that $C(R)$ behaves as a power of R for small R. By plotting $\log C(R)$ versus $\log R$ one can calculate D_2 from the slope of the curve. For an attractor, whose dimension is not known, it is necessary to calculate $C(R)$ for several embedding dimensions. In fact one should choose the maximal dimension n of the embedding phase-space twice the dimension of the attractor. The evaluation of the calculated correlation dimension D_2 should converge towards a saturation value. That means $\Delta D_2 / \Delta n = 0$ for some $n > n_0$.

3.2 Technical Considerations

The first step is to find the phase-space description of the digitized data. We have followed Taken's proposal (1981) and used the time-shift method. This

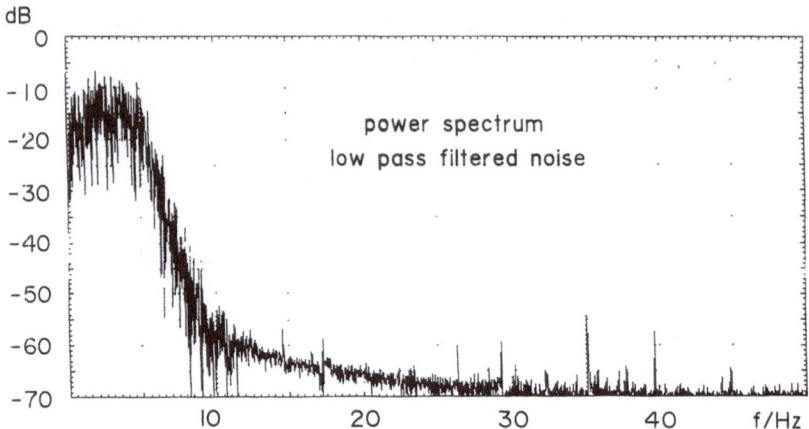

Fig. 6. Spectrum of a low pass filtered noise (cut-off frequency 5 Hz, 48 dB/octave)

means that a point x^i of the k-dimensional phase space is defined as

$$x^i = (x_i, x_i + \tau, \dots, x_i + (k-1)\tau)$$

where $x_s = x(s \cdot \Delta t)$ and Δt is the sampling interval.

Here, one of the most important questions is the choice of τ. In principle, the calculation of the dimension of an attractor should be independent of the choice of τ. In practice, this independence is given in an interval $\tau_1 < \tau < \tau_0$, where τ_0 is the first zero in the autocorrelation function.

In order to choose the adequate τ, we analyzed a low pass filtered noise, the spectrum of which is shown in Fig. 6. The calculation of the correlation integral was performed for $\tau = (10, 20, 30, 40, 50, 60, 70, 80, 90, 110, 130, 150 \text{ ms})$. Figure 7 illustrates the dependence of D_2 on different values of τ for three embedding dimensions ($n=2$, $n=5$, $n=7$). One can detect that for different embedding dimensions n, the higher the value of τ is, the higher the calculated correlation dimension D_2. It is obvious that the highest value of D_2 is determined by the embedding dimension n. But for a certain value of $\tau = \tau_{\min}$ no increase of D_2 is observed. In Fig. 8 a plot of $\tau_{\min}$ versus the embedding dimension n is illustrated. Obviously, for every n the maximal value of D_2 is calculated if $\tau > 90$ ms. The autocorrelation function analysis (ACF) of the low pass filtered noise shows that the first zero of the ACF is $\tau_0 = 90$ ms, too. This means that there should be no influence on the calculation of D_2 in a range $\tau_1 < \tau < \tau_0$.

In all of our computations, data sets of $N = 4096$ points have been used. In principle, it should be sufficient to consider only one point x^* of the attractor and evaluate the correlation integral $C(r)$ for all distances $|x^* - x^j|$. In our computations *every* distance $|x^i - x^j|$ ($i \neq j$) has been taken into account and the results have been averaged. The norm $\| \quad \|$ we used is the Euclidean norm

$$\| \quad \| = (\textstyle\sum (x_i - y_j)^2)^{1/2} \ .$$

Identical results will be obtained if the maximum norm is chosen

$$\| \quad \| = \max |x_i - y_j|$$

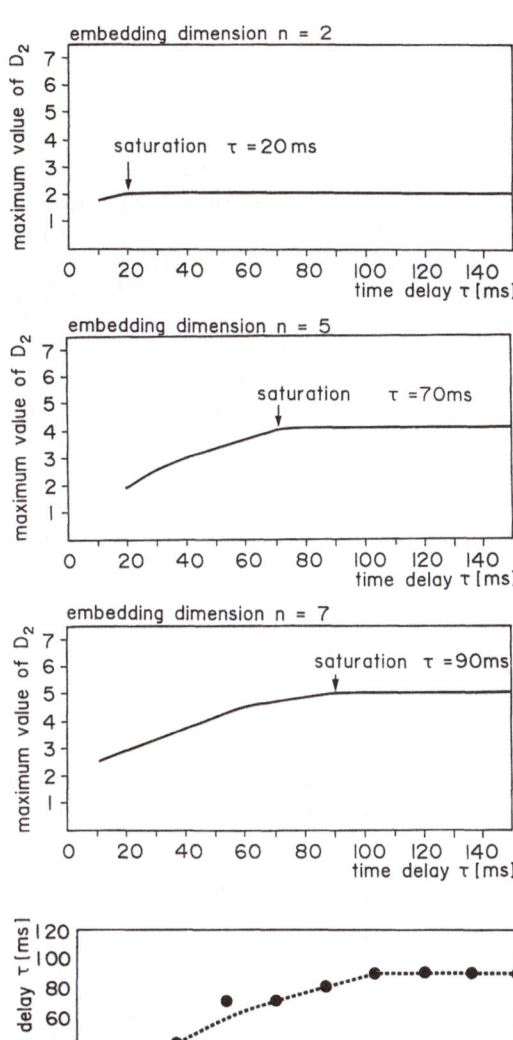

Fig. 7. For three embedding dimensions ($n=2$, $n=5$, $n=7$), the higher the value of τ, the higher the calculated correlation dimension D_2. It is evident that D_2 is lower than or equal to n. For a certain saturation value τ_{min}, no increase in the calculation of D_2 is observed

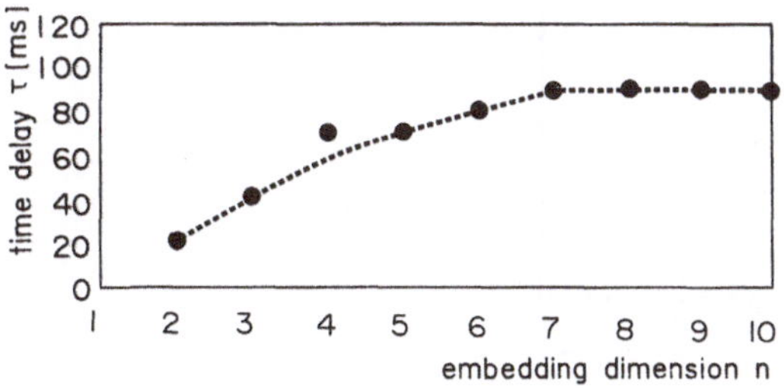

Fig. 8. The saturation value of τ_{min}, from which no further increase in the calculation of D_2 is observed, depending on the embedding dimension n

The slopes of the curves in the plot of "$C(r)$ versus log r" have been evaluated using the linear regression method. The logarithmic scale of the radii r was usually about a range of two decades, which was separated into 64 equal parts, though we evaluated the slope of the curves from 12–24 points. If the correlation coefficient of the linear regression was smaller than 0.98, the logarithmic scale of the radii was changed. Then the dimension D_2 was evaluated for a second time by separating a range of one decade or less into 64 equal parts.

4 Results

4.1 Early Results During Slow-Wave Sleep

In order to analyze the dimensionality of field potentials during slow-wave sleep (SWS), five cats with chronically implanted electrodes were studied. The chronic electrodes were needles of small exposed tips, implanted cortically (subdural) in the auditory cortex (GEA), and subcortically in the hippocampus (HI), and reticular formation (RF). The 15 experimental trials were recorded during SWS. The intracranial EEG signals were low-passed to 50 Hz and then digitized using a 12-bit AD converter and stored in the memory of an HP 1000-F computer. The sampling frequency was $f_s = 100$ Hz for all trials. Dimensions of the EEG signals were evaluated over a time period of about 20 s ($N = 2048$) and 40 s ($N = 4096$). Details of the software have been described elsewhere (Röschke 1986; Röschke and Başar 1988). The phase-space was constructed using the time-delayed coordinate proposed by Takens.

The evaluation of the correlation dimension for the auditory cortex is shown in Fig. 9, which represents an evaluation of the activity from the cat named "Toni" in SWS. One can detect that the slopes of the curves by plotting $N(r)$ versus log r converge against a saturation value of about $D_2 = 5.00$. The time lag τ was $\tau = 50$ ms. Identical results are obtained, if a time lag between 20 ms and 80 ms is chosen.

In order to underline the advantage of this procedure, the power spectrum of the EEG is illustrated (Fig. 10). This spectrum resembles that of colored noise. On the contrary, the case of the phase-space description and the evaluation of the correlation dimension lead to a convergence. Such a saturation is not reached in the

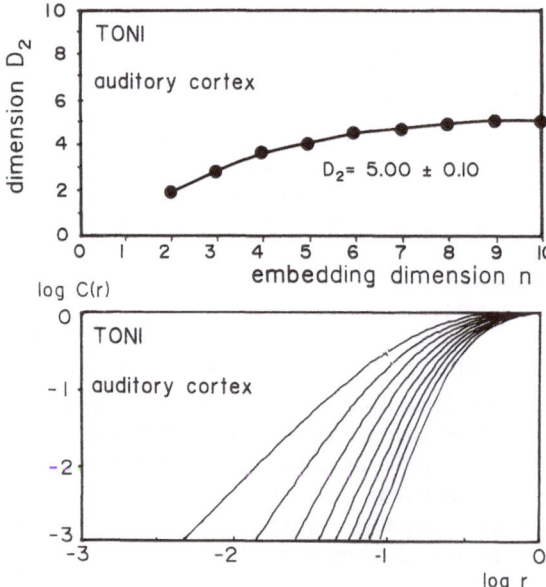

Fig. 9. The plot of log $C(r)$ versus log r (*at the bottom*) leads to a convergence of the slopes of the curves with a saturation value of $D_2 = 5.00$ (*at the top*); 4096 data points, $\tau = 50$ ms, cat "Toni," SWS, auditory cortex

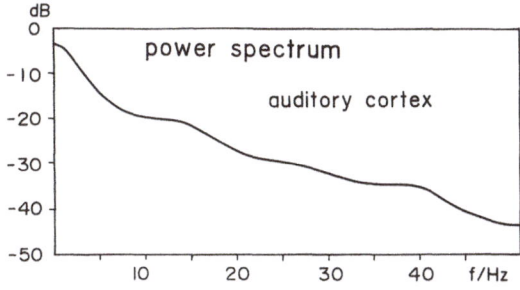

Fig. 10. Power spectrum of EEG epochs analyzed over a period of 40 s
(analyzed in Fig. 9). The spectrum was
smoothed by means of a previously
published linear prediction method
(Röschke 1986)

Table 1. Mean values of 15 trials (five cats) from the auditory cortex, reticular formation, and hippocampus

$$D_{GEA} = 5.06 \pm 0.31$$
$$D_{RF} = 4.58 \pm 0.38$$
$$D_{HI} = 4.37 \pm 0.36$$

case of a noisy signal. This is the difference between noise and strange attractors.

The mean values from 15 trials, each of 40 s measured in the acoustic cortex, the reticular formation, and the hippocampus of the cat brain of "Toni" during SWS, are shown in Table 1.

The standard deviations lie within the same range and are in all cases smaller than 10% of the mean. Significant differences between cortical and subcortical structures of the cat brain could be demonstrated. Detailed information on the results from five cats have been presented by Röschke and Başar (1988).

It is to be emphasized that the dimension D_2 of the investigated system, in our case the brain, varies from 4.37 in the hippocampus to up to nearly 5.00 in the cortex and that the dimensions of the different attractors are also relatively stable in experiments with individual cats. Moreover, in nearly 90% of the trials the maximal dimension is detected in the auditory cortex.

4.2 High Frequency Behavior of the Cat Cerebellar Cortex and Brain Stem

Although the spontaneous electrical activity of the brain which we call EEG has its most ample components in a frequency region between 1 and 50 Hz, it is well known that a number of higher frequency components can be observed in the field potentials of the brain.

Usually, the electroencephalographer turns the EEG low-pass filter down to 30 Hz routinely and only frequency regions of theta, alpha, and beta waves are analyzed or interpreted. It was a long-standing interest of our research group to try to describe the brain's spontaneous field potentials in the brain stem (reticular formation and inferior colliculus) and the cerebellum. In these structures we were able to demonstrate resonance phenomena and spontaneous field potential power

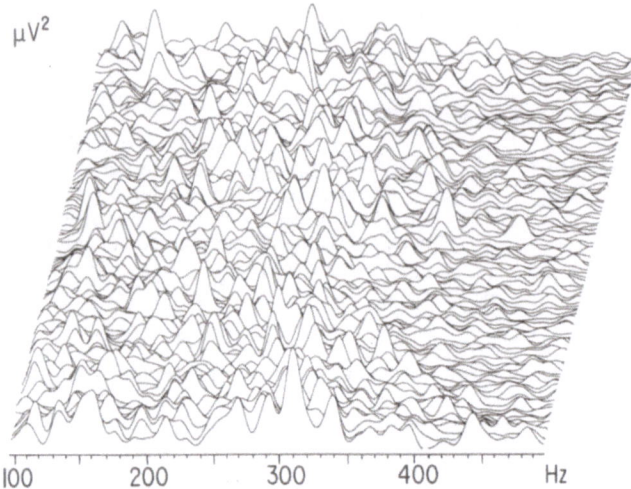

μV²

100   200   300   400   Hz

Fig. 11. Power spectral analysis of the electrical activity (cerebellar cortex) during waking stage from cat "Kati"

peaks in a frequency region between 100 and 1000 Hz (Başar 1980). Since the days of Adrian (1935) it has been well known that the cerebellum depicts a high frequency component between 180- and 300-Hz activity. Figure 11 shows a power spectral plot of the activity of the cerebellar cortex measured from a chronically implanted cat during waking stage. Spectral peaks in the frequency range over 100 Hz are observed in all the studied cats ($N=4$). Mostly, activity is dominant in the 200- to 300-Hz frequency range.

   One of our questions is how to consider the principle of self-similarity along the frequency scale of the brain's spontaneous activity. Is it possible to observe a self-similarity of brain waves in the higher frequency regions between 100 and 1000 Hz by comparing them with the lower frequency region of 1–100 Hz? According to this question we have undertaken the following evaluation: We translate the time scale, depending on the higher frequency, by a factor of 100. Time periods of 400 ms were, therefore, investigated, choosing a sampling frequency of 10 kHz and a low pass of 5 kHz. In order to eliminate the influence of the lower frequency components, the data have been high-pass filtered with a cut-off-frequency of 100 Hz. Field potentials of four freely moving cats with chronically implanted electrodes in the cerebellar cortex and the inferior colliculus were investigated during the waking stage. For the cerebellum a mean value of the correlation dimension of about $D_{CE}=7.05\pm0.15$ and for the inferior colliculus a mean value of $D_{IC}=6.70\pm0.20$ was obtained. Table 2 presents the correlaton dimension D_2 computed for all the experimental data from four cats and all experiments. In both cases an ensemble of 65 trials was investigated. But a convergence of the slopes, which is required to determine the correlation dimension, was observed only in approximately 25% of the studied data. In other words, in 75% of the investigated time periods the EEG signal cannot be distinguished from a noise. Only in 25% of the recordings can it be concluded that the EEG signal was a deterministic signal.

Table 2. Correlation dimension D_2 of high frequency components of ongoing activity in the cerebellum and inferior colliculus of several cats (100–5000 Hz) recorded from implanted semi-microelectrodes during the waking stage

Cerebellum

Cat	Successive samples of 400 ms						
Flic	7.14	7.01	7.14	7.08	7.14	7.14	
Maus	7.14	6.94	7.14	7.19	6.94	7.10	7.05
Kati	6.80	6.95					
Lucy	7.20	7.35	7.15	6.84	6.75		

Mean value: $D_2 = 7.05 \pm 0.15$

Inferior colliculus

Cat	Successive samples of 400 ms					
Flic	6.75	6.75	6.82	6.77	6.75	6.75
Maus	6.99	6.41	6.92			
Kati	6.57	6.57				
Lucy	6.60	6.66	6.50			

Mean value: $D_2 = 6.70 \pm 0.16$

The important difference from the results of SWS-EEG is this: For the SWS-EEG in nearly 75% of the results a convergence was observed, whereas for the waking state and high frequencies a real plateau is rarely found. In the waking state every movement artifact influenced the high frequency EEG recordings considerably; inevitable complications cannot be avoided by artifact rejection. It is important to note that the observations in the high frequency range depict a significant difference between the dimension of the cerebellum and the inferior colliculus. Even in this preliminary evaluation a difference between the two structures can be demonstrated.

If the cats fell into the SWS stage during the experiment, a reduction of the dimension of the cerebellar attractor to values near $D_{CE} = 6.80$ was observed. A systematic analysis for the inferior colliculus has not yet been undertaken.

We have to make a technical restriction on the interpretation of the results. Could the rejection of frequencies lower than 100 Hz by high-pass filtering influence our results concerning the strange attractor properties of the higher frequency EEG? In order to evaluate this point we have used a noise-generator (Wavetek-Rockland) and filtered the signals with a high-pass filter of 100 Hz (48 dB/octave). Figure 12 presents slopes of the curves by plotting $C(r)$ versus log r for the high-pass filtered noise in comparison with the cerebellar signal. For the noise no saturation of the slopes is observed. Embedding in successively higher dimensional phase-spaces, it was impossible to detect any subspace (with finite dimension) somehow describing the dynamics of the noise signal. Contrary to the cerebellar signal the high-pass filtered noise does not show convergence and accordingly no dimensionality. Therefore, we tentatively conclude that the filtering procedure does not drastically influence our results on attractor properties of the

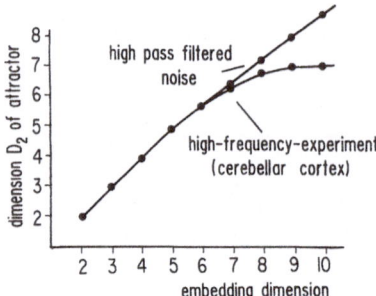

Fig. 12. The calculation of D_2 for a high pass filtered noise (high pass filter, 100 Hz) and the EEG signal from the cerebellar cortex. The EEG signal converges to a saturation value, whereas the noise which is filtered in the same frequency range does not show any convergence

cerebellum and inferior colliculus. Certainly the results indicate that the frequency region between 100–1000 Hz does not have properties of noise. More exact evaluation of D_2 requires more experiments and better mathematical and technical tools.

4.3   Are Transition Stages Also Reflected in the Correlation Dimension?

A survey of the correlation dimension D_2, which is described in contributions presented in this volume, shows a variability between 2 and 11. This variation depends on the state of the experimental subjects, on pathology, and on the evolutionary stage of the brains under study – and no doubt other factors as well. An important goal would be to interpret this range of values. As Freeman comments in the final discussion section "How Brains May Work" (Part V of this volume), it would be most useful to pay attention to the state changes, in other words, to bifurcations of brain activity rather than to merely classify the dimensionality. According to this view we have performed some experiments with freely moving cats and with human subjects who displayed abundant alpha activity with eyes closed.

Figure 13 shows the power spectra of the cat hippocampus with a dominant theta activity during the waking stage. The results indicate that if the cat brain state has a transition from waking stage to SWS stage, the maxima of power jumps down from about 5 Hz to between 1–3 Hz with several minor peaks. When the cat woke up, again a marked theta activity was observed and in this stage, D_2 was relatively high ($D_2 = 5.24$). During the SWS stage the dimension was reduced and had a value of $D_2 = 3.96$. Following the SWS stage D_2 again reached higher values of around 5.

Figure 14 illustrates the power spectra of the EEG activity of the subject J. K. who showed regular alpha activity with eyes closed. Here again it can be seen that the EEG activity reached slightly different states during the same recording session. In the present case it is impossible to describe exactly in which cognitive state the brain is. Elsewhere in this volume Başar et al. have described the same subject J. K. reaching long-standing alpha activity periods depicting marked time coherency depending on cognitive tasks; because of the short duration of the cog-

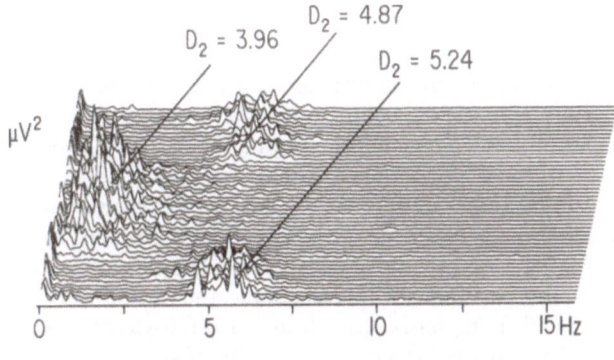

Fig. 13. Power spectra (5-s epochs) of the hippocampus (cat "Jenny") and correlation dimension D_2 (for 40-s epochs) during theta activity (waking stage) and slow-wave sleep

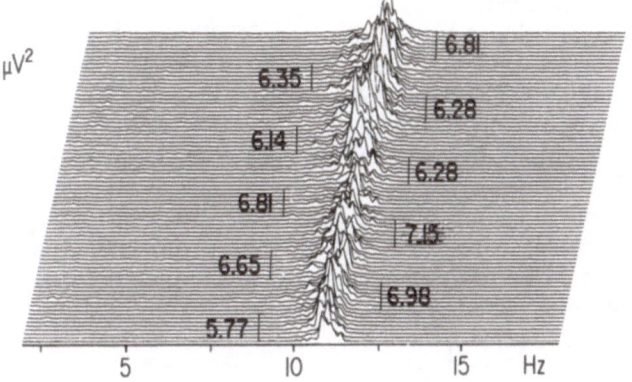

Fig. 14. Power spectra of EEG activity for 5-s epochs in subject J. K. (occipital recording) and correlation dimension D_2 for 40-s epochs

nitive tasks the evaluation of D_2 was not made. However, preliminary results indicate a trend towards a lower dimension during these brief tasks.

What conclusion is to be drawn from such a measurement where we observed a transient EEG activity? In this paper, we propose that together with a description of correlation dimensions one should use other parameters in parallel to D_2, for example, the power spectra which describe not only the overall power but also the quality of EEG signals in terms of its several frequency components. Another descriptor consists in estimating the RMS values in the conventional EEG frequency bands. Other important points to be analyzed are differences between closed eyes and opened eyes experiments, the range of filters and of EEG amplifiers, and also the classification of subjects as "alpha type" or "beta type". In the EEG studies another important but neglected point is the amount of 40-Hz activity which can reach maximal amplitudes of 10 to 14 µV depending on experimental conditions (Başar et al. 1987). This 40-Hz frequency is mostly masked in the conventional EEG analysis by visual inspection and also in the power spectral analysis. Digital filters can be used to describe the trend of this activity approximately. We do not yet know what the contribution of 40-Hz activity is to the EEG dimensionality. Only after clearly defining the experimental conditions and the distribution of the EEG in various stages, can use of the correlation dimension be valuable as a descriptor which combines and extends some other important properties of brain waves.

5   Discussion

Our research group's interest in the nonlinear analysis of the EEG dates back to 1972. Başar (1972) described a simple model of nonlinear interactions in the spontaneous activity of the brain. This model was a generalization of Erich von Holst's analysis of relative coordination in the CNS. Later, this tentative model concept was extended, by mentioning the Duffing equation, to describe evoked potentials as deterministic forced oscillators and EEG as randomly forced oscillations (Başar et al. 1979; Başar 1980, 1983). Başar and Röschke (1983) assumed that EEG components in several frequencies could be manifested as part solutions of Navier-Stokes equations. Shortly after the pioneering study of Babloyantz et al. (1985), Röschke and Başar (1985 b) reported the dimensionality of substructures of the cat brain during slow-wave sleep.

The implications of the expression "strange attractor EEG" have been described in an increasing number of studies (Röschke and Başar 1988; Başar et al. 1988; Babloyantz 1988; Freeman 1988; Başar 1983, 1988, Epilogue). We also want to refer to the Epilogue written by Bullock and other contributions in the present volume (see Rössler, Freeman, Babloyantz, Saermark et al., Graf and Elbert, this volume).

In this discussion we will present some remarks concerning the new data and make some comments regarding future research. The ensemble of data presented here might help to elucidate in part a few points that still remain obscure in this important new trend to describe the dimensionality of EEG.

5.1   The Variability of D_2 in the Human EEG

The description of dimensionality of the human EEG continues to show enormous fluctuations in studies by several authors (see Babloyantz, Graf and Elbert, Saermark, this volume). The correlation dimension of human EEGs varies between 5 and 11 in analyses so far known to us.

In the present study we reported that during a measuring period of 25 min the correlation dimension of the EEG of the same subject varies between 6 and 7, more than 20% during a short period. It is perhaps understandable that D_2 could show larger differences between subjects, during undefined states, with various electrode locations, digital filters, and types of EEG recording (50-Hz noise, surrounding, sound-proof room). We highly recommend the use of standard experimental conditions taking into account all the above criteria. At least investigators should describe experimental conditions exactly. This is important, since the serious use of D_2 in future studies dealing with the observation of fluctuations (transition changes) can be enormously useful.

5.2   Bifurcations

In Section 4 of the present study it is illustrated that if a cat goes from the slow-wave sleep to the waking state the power spectral peaks shift from delta activity to theta activity and the correlation dimension increases from about 4 to 5. This is a consistent observation of the hippocampal activity where changes of D_2 are always well correlated with the delta→theta transition. If the dimension D_2 is an important measure to describe states of the brain, the determination of transitions (or bifurcations) by using D_2 as an indicator would gain importance for studies of behavior of the CNS. We mention a word of caution for this type of description. The correlation dimension D_2 of the hippocampus during the waking state and D_2 of the auditory cortex during SWS are approximately 5. Accordingly, there is no difference between these two different brain states. Therefore, it would be extremely inadequate to consider D_2 as a universal indicator of the brain state. It provides experimenters, for the time being, with an abstract quantification which can be helpful in connection with several other biological descriptors of the brain state.

5.3   Power Spectra

The use of power spectra and digital filtering of the EEG seems to be a useful mathematical tool to complement the interpretation of the correlation dimension D_2. As indicated above (Sect. 5.2) the dimension of the cortex during sleep as well as wakefulness and of hippocampus during the waking stage give the same values of $D_2 = 5$. In that case, spectral analysis enables the experimenter to classify the brain waves. If the dimension of a human EEG during the waking stage is $D_2 = 6$, this might be the result of a mixture of theta, delta, and alpha waves and 40-Hz activity. We do not have models which describe, step by step, the distributions of spectral peaks, their weights, and the resulting correlation number. The use of power spectra might also help to reconcile discrepancies between results of various research groups, especially from records during the waking stage.

5.4   High Frequency Activity of the Cerebellum and Brain Stem

The high frequency spectral activity of the cerebellum and inferior colliculus have been mostly neglected by investigators. The experimental data described in this report demonstrate that in a frequency range between 100 Hz and 1000 Hz the brain has a high frequency attractor which is not as stable as the lower frequency attractor during SWS. A comparison with a noise signal which has been filtered with the same high-pass digital filters showed a completely different behavior. Therefore, it can be concluded that the high frequency cerebellar and collicular activity are not due to noisy signals but most probably stem from deterministic activities.

We want to mention that the brain's electrical activity might show self-similarity comparing the attractors in frequency regions of 1–100 Hz and 100–1000 Hz, although the evidence is still no more than suggestive. We would echo Adey's comments on "Bioinstrumentation: cutting edge and limiting factor in the future of brain research (this volume) concerning the need for looking to higher frequency windows.

From the mathematical viewpoint we do not assume or demonstrate the existence of two strange attractors in the frequency regions of 1–100 Hz and of 100–1000 Hz. For the time being the analysis undertaken by considering two frequency windows is only an approximation of the most appropriate analysis. The better procedure would be to analyze only one window between 1–1000 Hz (or higher). Unfortunately, this is still technically impossible due to computer limitations. Accordingly, the analysis in the higher frequency window has to be considered for the present as an indicator that the high frequency activities also point to deterministic chaos and that the activity between 100 and 1000 Hz is not simply noise.

The use of the brain state matrix concept which is proposed by Başar (Part V of this book) including several linear and nonlinear indicators seems necessary to us in studies related to the evaluation of D_2 according to the description given in Section 5.3.

6   Summary

We describe and compare the correlation dimension (D_2) of EEG activity during waking and sleep stages in various intracranial structures of the cat brain and in scalp activity of the human brain. D_2 is evaluated using Grassberger and Procaccia's algorithm. The following results were obtained: (1) During slow-wave sleep, D_2 has values between 4 and 5 depending on the brain structure under study; highest values were recorded in the auditory cortex (D_2 ca. 5) and lowest in the hippocampus (D_2 ca 4). (2) When the hippocampal EEG of a sleeping cat shows a transition from slow-wave sleep to the waking stage with dominant hippocampal theta waves, D_2 jumps from a value of approximately 4 to 5. (3) The cerebellar cortex and inferior colliculus of the cat brain depict a less stable strange attractor in the high frequency range between 100 and 1000 Hz (D_2 varies between 6 and 8). (4) During a recording session with a human subject showing abundant alpha activity, the D_2 is likely to show large fluctuations within a short period; in the example presented D_2 fluctuated between 6 and 7. (5) EEG activity gives a highly different result from that of white noise. (6) Power spectra of the EEG is a necessary and complementary brain descriptor to D_2. The correlation dimension can be more valuably used as one of several descriptors of the brain state; using only the correlation dimension to describe the brain state can lead to misinterpretations.

References

Abraham RH, Shaw CD (1983) Dynamics – the geometry of behavior. Part 2: chaotic behavior. Aerial Press, Santa Cruz

Adrian ED (1935) Discharge frequencies in cerebral and cerebellar cortex. J Physiol 83:32–33

Babloyantz A (1988) Chaotic dynamics in brain activity. In: Başar E (ed) Dynamics of sensory and cognitive processing by the brain. Springer, Berlin Heidelberg New York, pp 196–202 (Springer series in brain dynamics, vol 1)

Babloyantz A, Nicolis C, Salazar M (1985) Evidence of chaotic dynamics. Phys Lett[A]:152–156

Başar E (1972) Remarks on mathematical signal processing by the brain during rhythmic neuro-physiological stimulation. Int J Neurosci 4:71–76

Başar E (1980) EEG-brain dynamics. Relation between EEG and brain evoked potentials. Elsevier/North-Holland, Amsterdam

Başar E (1983) Toward a physical approach to integrative physiology. I. Brain dynamics and physical causality. Am J Physiol 245(4):R510–R533

Başar E (1988) EEG-dynamics and evoked potentials in sensory and cognitive processing by the brain. In: Başar E (ed) Dynamics of sensory and cognitive processing by the brain. Springer, Berlin Heidelberg New York, pp 30–55 (Springer series in brain dynamics, vol 1)

Başar E, Röschke J (1983) Synergetics of neuronal populations. A survey on experiments. In: Başar E, Flohr H, Haken H, Mandell AJ (eds) Synergetics of the brain. Springer, Berlin Heidelberg New York, pp 199–200 (Springer series in synergetics, vol 23)

Başar E, Demir N, Gönder A, Ungan P (1979) Combined dynamics of EEG and evoked potentials. I. Studies of simultaneously recorded EEG-EPograms in the auditory pathway, reticular formation and hippocampus of the cat brain during the waking stage. Biol Cybern 34:1–19

Başar E, Rosen B, Başar-Eroglu C, Greitschus F (1987) The associations between 40 Hz-EEG and the middle latency response of the auditory evoked potentials. Int J Neurosci 33:103–117

Başar E, Başar-Eroglu C, Röschke J (1988) Do coherent patterns of the strange attractor EEG reflect sensory-cognitive states of the brain? In: Markus M, Müller S, Nicolis G (eds) From chemical to biological organization. Springer, Berlin Heidelberg New York

Freeman WJ (1988) Nonlinear neural dynamics in olfaction as a model for cognition. In: Başar E (ed) Dynamics of sensory and cognitive processing by the brain. Springer, Berlin Heidelberg New York, pp 19–29 (Springer series in brain dynamics, vol 1)

Grassberger P, Procaccia I (1983) Measuring the strangeness of strange attractors. Physica [D]9:183–208

Lorenz EN (1963) Deterministic nonperiodic flow. Atmos Sci 20:130

Röschke J (1986) Eine Analyse der nichtlinearen EEG-Dynamik. Doctoral dissertation, University of Göttingen

Röschke J, Başar E (1985a) A phase portrait analysis of the EEG and evoked potentials. Electroencephalogr Clin Neurophysiol 61:S114

Röschke J, Başar E (1985b) Is EEG a simple noise or a "strange attractor"? Pflugers Arch 405(2):R45

Röschke J, Başar E (1988) The EEG is not a simple noise. Strange attractors in intracranial structures. In: Başar E (ed) Dynamics of sensory and cognitive processing by the brain. Springer, Berlin Heidelberg New York, pp 203–216 (Springer series in brain dynamics, vol 1)

Schroeder MR (1986) Number theory in science and communication. Springer, Berlin Heidelberg New York

Schuster HG (1984) Determnistic chaos. An introduction. Physik, Weinheim

Takens F (1981) Detecting strange attractants in turbulence. In: Rand DA, Young LS (eds) Dynamical systems and turbulence, Warwick 1980. Springer, Berlin Heidelberg New York, pp 366–381 (Lecture notes in mathematics, vol 898)

Magnetoencephalography and Attractor Dimension: Normal Subjects and Epileptic Patients *

K. Saermark, J. Lebech, C. K. Bak, and A. Sabers

1 Introduction

In recent years methods from nonlinear dynamics analysis have been applied to examine the human electroencephalogram (EEG) with respect to attractor dimensions, Lyapunov exponents, etc.; using the same methods, Röschke and Başar (1988) have performed an analysis of EEG activity in cortical and subcortical structures of chronically implanted cats with respect to the fractal dimension of an EEG attractor during slow wave sleep. For the human EEG, Babloyantz et al. (1985) were able to determine an attractor dimension for two sleep stages (stages 2 and 4) whereas for REM sleep and for the awake state no attractor dimension was quoted. The dimensions found for the two sleep stages were close to 5 (stage 2) and 4 (stage 4). The analysis was based on data for two and three subjects, respectively, the experimental data consisting of time series of 4000 points with a sampling time of 10 ms. Similarly, for the chronically implanted cats (electrode positions: acoustic cortex, hippocampus, and reticular formation), Röschke and Başar quote a value of 4–5 for the EEG attractor during slow wave sleep. Further works by Babloyantz and Destexhe (1986) and Layne et al. (1985) quote still other values for attractor dimensions obtained under a number of conditions: anesthetized subjects, epileptic patients, alpha traces, etc.

In analyzing the EEG data for the attractor – correlation dimension the above references make use of the Grassberger-Procaccia algorithm (Grassberger and Procaccia 1983). Here one calculates the correlation function, $C(r)$, as defined in the above references and examines its behavior for small values of r. $C(r)$ behaves like $\sim r^d$ and if a plot of d versus the embedding dimension n shows a saturation value for d, then this value is taken as the correlation dimension for the attractor. It can be argued, however, that the Grassberger-Procaccia algorithm is not ideally suited for the determination of the dimension of high-dimensional attractors. Thus Layne et al. (1985) discussed problems associated with dimensional analysis of EEG data (T3-C3 and P3-O1 electrodes) and found – even for long data sets containing 15000 points – that the dimension could be ascertained only with a very high degree of uncertainty (of the order of 50%–100%). Based on calculations for shorter subsections (4000 data points) they also showed that the dimension depends on the number of data points used in the analysis and on the position of the subsection of data points within the larger set of data points. Hence their conclusion "The EEG is both non-stationary and high-dimensional. Therefore, we cannot speak of dimension in the strict sense. Our calculations of 'dimen-

* Originally published in Başar E, Bullock TH (eds) Brain dynamics. Springer, Berlin Heidelberg New York, pp 149–157 (Springer series in brain dynamics, vol 2). Cross references refer to that volume.

sion' have a meaning only in a comparative sense," is well taken and the word "dimension" could appropriately be replaced by their "apparent dimension" or "dimensional complexity." However, even with this caution in mind it still appears very tempting to use the Grassberger-Procaccia algorithm in the determination of correlation dimensions of EEG and magnetoencephalographic (MEG) data. This is especially so since there are very few other methods for performing "global" analyses of EEG and MEG recordings. It should be remarked, however, that the criticism with respect to nonstationarity is especially significant for EEG and MEG recordings for epileptic patients. Such recordings are normally inter-ictal recordings and the interictal state is in general not a stationary state. A dimensional analysis of MEG recordings was briefly presented previously (Saermark 1986). In this paper we present a number of results for normal subjects and for epileptic patients. In particular we point out that the present experimental data indicate that the correlation function $C(r)$ – as calculated in the Grassberger-Procaccia algorithm – for epileptic patients shows a behavior which is different from that shown for normal subjects.

2 Results

The MEG recordings discussed in this paper were obtained for two groups, one population consisting of five normal subjects (two female, three male; age 21–25 years) and another population consisting of four epileptic patients (all female; age 20–29 years).

The magnetic field was measured by means of a first order SQUID gradio-meter (BTI) located in magnetically quiet surroundings. The base line of the gradiometer is 6.3 cm, and the lower sensing coil (diameter 2.4 cm) is situated 1.2 cm above the bottom of the cryostat tail. The latter was separated from the subject's skull by a distance of 1–2 mm. Measurements of the magnetic field com-ponent normal to the skull were performed for both right and left hemisphere (RH, LH) positions. Each MEG recording consisted of 8192 points with a sam-pling time of 10 ms, yielding a total recording epoch of nearly 82 s. For each nor-mal subject the MEG was recorded in 10–15 positions, all of which lay above the temporal lobe or the immediately adjacent areas of the brain. For each of the epi-leptic patients 50–60 MEG recordings were taken at positions evenly distributed across the hemisphere suspected for epileptic foci, with an additional 10–15 re-cordings being taken at the opposite hemisphere.

In analyzing the MEG data for "attractor dimension" we used the Grass-berger-Procaccia procedure and thus calculated the correlation function $C(r)$ as a function of r for a number of embedding dimensions: $n = 2, 4, 6, \ldots 20$. Gener-ally 4000 data points were used in the analysis; however, in a number of cases this was increased to 8000 data points. For each subject only two or three measuring positions were used in order to keep the computer expenses within reasonable limits. For normal subjects positions close to the auditory cortex were chosen while for epileptic patients positions where the recordings showed a clear promi-nence of low-frequency waves (3–6 Hz) were chosen.

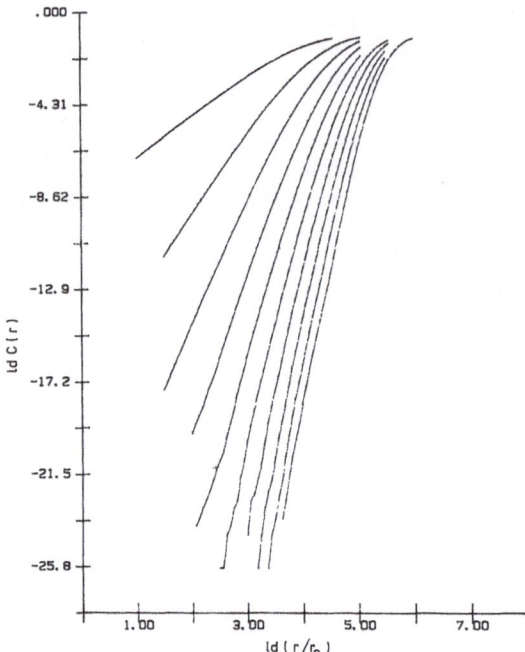

Fig. 1. The correlation function $C(r)$ for a normal subject based on MEG data. *Abscissa:* $\log_2 (r/r_0) \equiv ld (r/r_0)$, where r_0 is a reference r value. *Ordinate: ld C(r)*. Embedding dimensions 2, 4, 6, ... 20

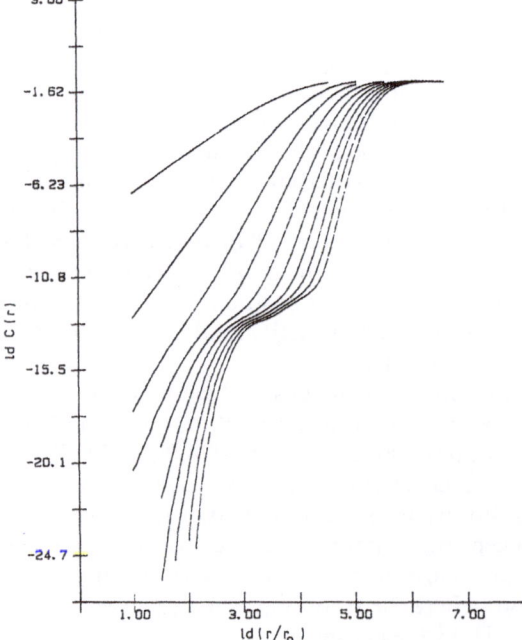

Fig. 2. The correlation function $C(r)$ for interictal MEG data from an epileptic patient. Axes and embedding dimensions as in Fig. 1

In Fig. 1 we show a plot of $\log_2 C(r)$ versus $\log_2(r/r_0)$, r_0 being a reference value for r, for a normal subject and for an MEG recording. We note that this subject showed particularly well developed alpha waves with maximal amplitudes of the order of 2500–3000 fT. For all of the embedding dimensions used, the curves for $C(r)$ show a linear portion with a slope which converges to a limiting value. For epileptic patients, on the other hand, $C(r)$ behaves quite differently, as is clearly seen in Fig. 2. For small values of the embedding dimension, say $n < 6$, the curves do show a linear behavior; however, for increasing values of the embedding dimension an inflection point, with a decreasing value of the slope of the tangent, gradually develops for abscissa values around 2–3. It follows that in this region of n values the curves show no overall linear behavior, but can be characterized by the presence of two inflection points: the marked one just mentioned and in addition one occurring at a larger r value. Further, at an r value smaller than the one for the marked inflection point, the tangent to the $C(r)$ curve shows a maximum slope slightly greater than the slope of the tangent at the larger r value. The behavior of $C(r)$ shown in Fig. 1 (normal subject) was consistently seen for all normal subjects (including a third group of five normal subjects for whom the analysis was not carried through in a systematic way), while the behavior of $C(r)$ shown in Fig. 2 (epileptic patient) was consistently seen for all epileptic patients. In comparing Figs. 1 and 2 it should be emphasized that the measuring instrument and the measurement procedure were identical for the two groups of subjects and that the raw data were used in the analysis, i.e., no filtering or other data handling was carried out before the dimensional analysis was performed. We therefore believe, until evidence is available to the contrary, that the difference in the behavior of $C(r)$ for the two groups of subjects is a genuine difference. To bring out this difference clearly, Fig. 3 shows $C(r)$ for an epileptic patient ($n = 18$) and for a normal subject ($n = 20$).

For the group of normal subjects we may plot the slope d of the linear region of $C(r)$ versus the embedding dimension, whereas for the group of epileptic patients this cannot be done. We have here chosen to plot the slope of the tangent at the inflection point occurring at the larger r value and also the slope of the tangent at the marked inflection point. In Fig. 4 the crosses indicate the population average for the group of normal subjects while the rhombohedral symbols show the results for a single epileptic patient. The vertically hatched symbols refer to an analysis based on MEG recordings (4000 data points) while the open symbols refer to an analysis based on EEG recordings taken 3 months after the MEG recordings (again based on 4000 data points). The cross-hatched symbol (for $n = 20$) refers to an analysis based on MEG recordings and 8000 data points. As the analysis of the EEG data also gave rise to the presence of the marked inflection point, in this figure we have chosen to show the results from the analysis of the EEG data, i.e., the horizontally hatched rhombohedral symbols which virtually coincide with the results from the analysis of the MEG data. From Fig. 4 one notes, first of all, that there is good agreement between the analysis of the MEG data and the EEG data, with respect to both the slope of the tangent at the upper inflection point and the slope of the tangent as the marked inflection point; further, for the epileptic patient there may be an apparent dimension of around 7, while the data for the population of normal subjects may indicate an apparent dimen-

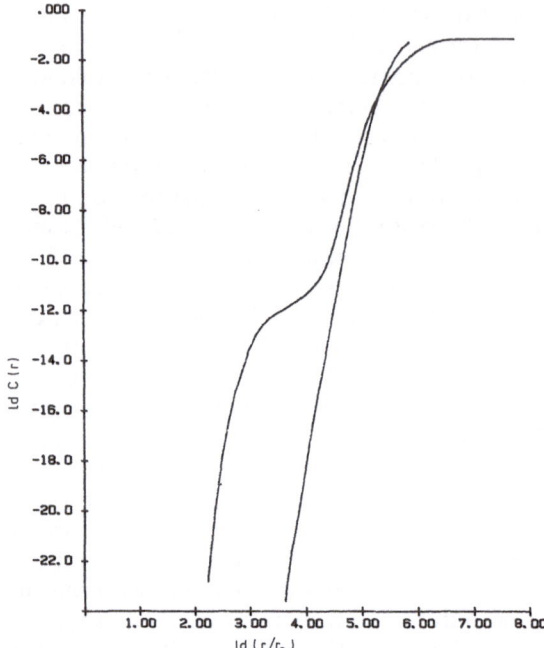

Fig. 3. Comparison of correlation function $C(r)$ for a normal subject ($n=20$) and an epileptic patient ($n=18$). The marked inflection point is visible for the epileptic patient, but is not present for the normal subject

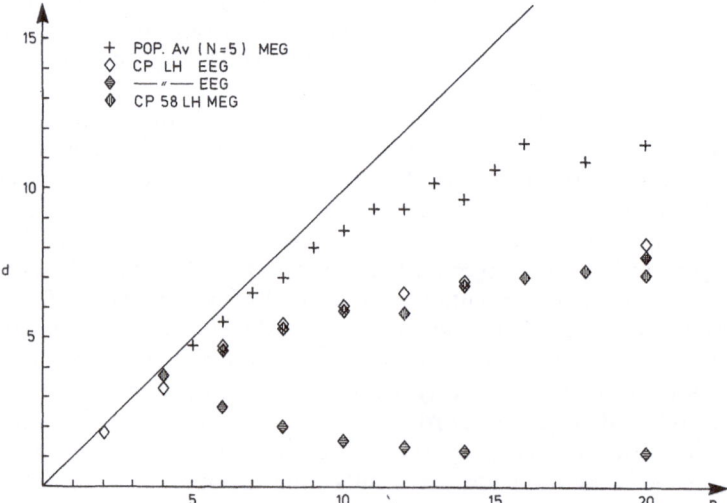

Fig. 4. Plot of the exponent d (see text) versus the embedding dimension for a population of normal subjects and for an epileptic patient CP. For the latter both MEG and EEG data have been used. See text for explanation of symbols

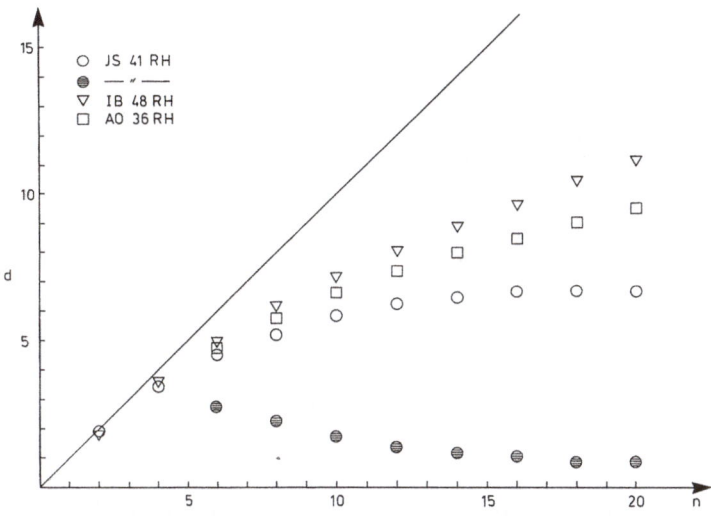

Fig. 5. Plot of the exponent *d* (see text) versus the embedding dimension for three epileptic patients (RH measuring position)

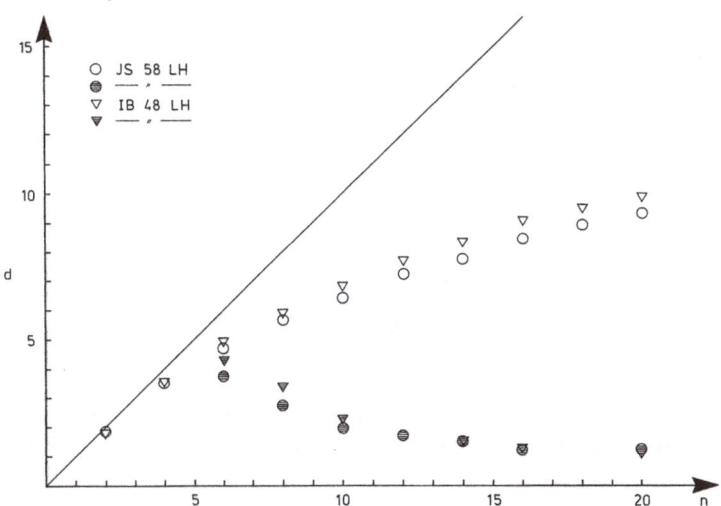

Fig. 6. Plot of the exponent *d* (see text) versus the embedding dimension for two epileptic patients (LH measuring position)

sion of around 11. Finally, the tangent slope at the marked inflection point approaches, rather rapidly, a value of 1 for increasing embedding dimensions; thus a bifurcation in *d* versus *n* appears to occur for an embedding dimensions of around 5–6 in the case of epileptic patients but not in the case of normal subjects.

In Figs. 5 and 6 we show results obtained for the three other epileptic patients considered in this paper. For patient JS one notes an apparent dimension between

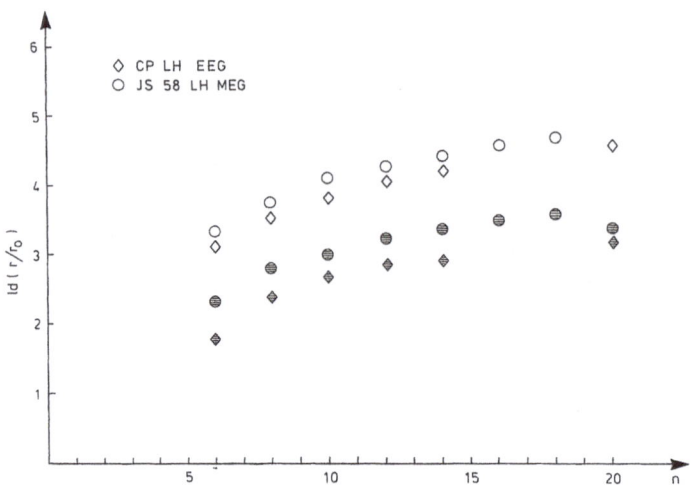

Fig. 7. Plot of the r value for the position of the marked inflection point (*open symbols*) and the upper inflection point (*hatched symbols*) versus embedding dimension for two epileptic patients. Note the nearly constant difference between the two positions

6 and 7 for the RH measuring position (Fig. 5) whereas the LH measuring position (Fig. 6) does not give rise to an apparent dimension. The latter is true also for the two remaining patients IB and AO (Fig. 5, 6). In all cases, however, there is a bifurcation point – the position of which may vary slightly from patient to patient – and in all cases the slope of the tangent at the marked inflection point approaches 1 for increasing values of the embedding dimension.

Finally, in Fig. 7 we show the r values for the position of the two inflection points. For the sake of clarity we show only data for two patients; however, the remaining two patients showed identical behavior. For the patient JS we have used MEG data and for the patient CP, EEG data. One notes a clear tendency toward a constant ordinate difference between corresponding open and hatched symbols. As the ordinate is a logarithmic scale this implies a constant ratio (slightly larger than 2) between the r values for the position of the upper inflection point and the marked inflection point. This was also found for the two other epileptic patients.

3   Discussion

We have here presented a number of results from a dimensional analysis of MEG data and some EEG data. In view of the cautioning remarks concerning the applicability of the Grassberger-Procaccia algorithm (see Sect. 1) with respect to the determination of the "dimension" of high-dimensional attractors, however, all conclusions must be very tentative. It is especially disturbing that Layne et al. [4] estimate the uncertainty in the dimension d to be as large as 50%–100%. On the other hand, in the experiments by Röschke and Başar (1988), where five chronically implanted cats were examined, the intersubject variation in d was considerably lower, i.e., of the order of 10%. Similarly, for the group of normal subjects considered here the intersubject variation in d for a given value of the embedding dimension n is also considerably smaller: less than 20%. From this point of view one could with some justification argue that the data for normal subjects may indicate an apparent dimension of the order of 11.

For the group of epileptic patients a surprising feature emerged in the correlation function $C(r)$, in the form of what we here denoted as "a marked inflection point" (see Figs. 2, 3). As a consequence one finds a bifurcation point in a plot of d versus n, as shown in Figs. 4–6; however, this bifurcation is only observed for the group of epileptic patients, not for the group of normal subjects. The slope of the tangent at the marked inflection point rapidly approaches the value 1 for increasing embedding dimension for all epileptic patients, whereas the slope of the tangent at the upper inflection point (see Figs. 4–6), as a function of the embedding dimension, behaves differently from patient to patient. In two cases (Fig. 4, CP 58 LH MEG; Fig. 5, JS 41 RH MEG) one does find an apparent dimension. In these two cases, visual inspection of the MEG recordings shows the presence of prominent low-frequency waves (mainly 3- to 5-Hz waves) for the whole time epoch; these records may therefore be considered nearly stationary and the finding of an apparent dimension be ascribed some significance. It is also noteworthy that for the patient CP (see Fig. 4) the dimensional analysis of EEG data – which, as mentioned earlier, were taken 3 months later than the MEG data – was in excellent agreement with the analysis of the MEG data. For the three remaining patients no EEG data were available to us; thus we cannot claim that the two types of data will always yield an identical apparent dimension. In this connection it is to be remarked that EEG data depend on the existence of volume currents whereas the MEG data are predominantly determined by intracellular currents. It is therefore not obvious that the two types of data should yield the same apparent dimension. For the three remaining epileptic patients one cannot claim the existence of even an apparent dimension (see Figs. 5, 6). It is a characteristic feature, however, that for the epileptic patients all points fall considerably below the points representative for the group of normal subjects (see Fig. 4), which could indicate that an apparent dimension for epileptic patients will be lower than for a normal subject, but, at the same time, that the present interictally recorded MEG data represent a nonstationary state. In this paper we have chosen to perform the dimensional analysis on the basis of the raw experimental data; it is conceivable, however, that by first applying bandpass filtering (e.g., 2–8 Hz) one might find an apparent dimension for all the epileptic patients.

The present data lead, as pointed out earlier, to the existence of a bifurcation point in a plot of d versus n. This could possibly be interpreted in the following way. For sufficiently large values of the embedding dimension, say $n > 6$, the phase-space points show a tendency to fall into two separate groups of a certain extent and separated by a certain distance. If this is the case, there will be a range of r values where $C(r)$ will show a tendency to flatten out, then (for increasing values of r) start to increase rapidly again before ultimately bending over and becoming constant for r going to infinity (in the logarithmic scale used in Figs. 1–3 the constant value is -1). The fact that (see Fig. 7) the r values for the positions of the two inflection points both appear to approach a limiting value for increasing values of n and the fact that there appears to be a constant ratio between these r values, i.e., a ratio nearly independent of the embedding dimension for $n > 6$, may support this hypothesis. We emphasize that this ratio appears to be nearly the same for all the epileptic patients examined in this study. It is, of course, quite a different matter to find out what are the neurological implications of such a hypothesis.

4 Summary

In this paper we present results obtained by applying the Grassberger-Procaccia algorithm to a dimensional analysis of MEG and in part also EEG data for a group a five normal subjects and for a group of four epileptic patients. On the basis of the data presented, it is concluded that the correlation function $C(r)$, as defined by Grassberger and Procaccia, behaves differently for the two groups. For one epileptic patient a comparison between an analysis based on MEG data and an analysis based on EEG data is performed, the comparison showing an excellent agreement. In a number of cases the analyses show a trend to an "apparent dimension," whereas in other cases they indicate a nonstationarity of the recorded experimental time series.

References

Babloyantz A (1988) Chaotic dynamics in brain activity. In: Başar E (ed) Dynamics of sensory and cognitive processing of the brain. Springer, Berlin Heidelberg New York

Babloyantz A, Destexhe A (1986) Low-dimensional chaos in an instance of epilepsy. Proc Natl Acad Sci USA 83:3513–3517

Babloyantz A, Nicolis C, Salazar M (1985) Evidence of chaotic dynamics of brain activity during the sleep cycle. Phys Lett [A]111:152–156

Grassberger P, Procaccia I (1983) Measuring the strangeness of strange attractors. Physica [D]9:183–208

Layne SP, Mayer-Kress G, Holzfuss J (1985) Problems associated with dimensional analysis of electroencephalogram data. In: Mayer-Kress G (ed) Dimensions and entropies in chaotic systems. Springer, Berlin Heidelberg New York

Röschke J, Başar E (1988) The EEG is not a simple noise: strange attractors in intracranial structures. In: Başar E (ed) Dynamics of sensory and cognitive processing by the brain. Springer, Berlin Heidelberg New York, pp 203–216 (Springer series in brain dynamics, vol 2)

Saermark K (1986) On physical models of elementary neural sources. International Workshop an Functional Localization: A Challenge for Biomagnetism. Sept 1–6, Torino

Chaotic Attractors in a Model of Neocortex: Dimensionalities of Olfactory Bulb Surface Potentials Are Spatially Uniform and Event Related*

J. E. SKINNER, J. L. MARTIN, C. E. LANDISMAN, M. M. MOMMER, K. FULTON, M. MITRA, W. D. BURTON, and B. SALTZBERG

1 Introduction

The olfactory bulb has both the cell types and the neurochemicals intrinsic to neocortex, but it has a much simpler and better understood neurophysiological structure (Shepherd 1970). The amplitude of the field potential that occurs on the top of each columnar unit is linearly related to the firing probability of the immediately underlying output cell (Freeman and Schneider 1982; Gray et al. 1984, 1986). Thus the recording of all such surface potentials, at the spatial frequency of the functional units, makes possible knowledge of the total output of the bulb without having to make massive microelectrode penetrations.

The amplitudes of these simultaneously recorded surface potentials form a pattern on the bulbar surface that is altered by a novel or classically conditioned odor (Freeman and Schneider 1982). This alteration is prevented by interstitial beta-receptor antagonists (Gray et al. 1984, 1986). Beta-receptor antagonism also blocks synaptic potentiation in hippocampal cortex (Bliss et al. 1983; Hopkins and Johnston 1984), an alteration which is thought to be a model for learning and memory (Anderson and Wigstrom 1980; Swanson et al. 1982). Thus the study of the spatial patterns on the bulbar surface (the output) during controlled stimulus events (the input) may lead to an understanding of how information is acquired and stored in a neural system composed of large numbers of neurons.

The learning-dependent changes in the surface patterns, although statistically significantly different during pre- and postodor intervals, are quite noisy in appearance (Gray et al. 1986). Such complex signals suggest an underlying chaotic dynamical process (Farmer 1982). Other investigators have reported fractional dimensionalities of surface potentials from neocortex (Babloyantz and Destexhe 1986, 1987; Röschke and Başar, this volume; Graf and Elbert, this volume). Our present data indicate that: (a) a single, global chaotic attractor may represent the total bulbar activity spatially, (b) an odor causes this widespread process to shift to a new attractor, and (c) the shift is to an attractor with a larger Hausdorff number if the odor is novel and a lower one if the odor is familiar.

2 Methods

Rabbits were surgically prepared, instrumented with an 8×8 electrode array on the lateral surface of the olfactory bulb, and behaviorally trained according to a

* Originally published in Başar E, Bullock TH (eds) Brain dynamics. Springer, Berlin Heidelberg New York, pp 158–173 (Springer series in brain dynamics, vol 2). Cross references refer to that volume.

previously described protocol (Gray et al. 1986). The spacing between each electrode was at the spatial frequency of the surface activity (500 μm). The bandpass (3 dB) was set to 20–200 Hz and the digitizing rate (12 bits) set to 500 kHz with 640 samples/second per channel. The bandpass eliminates the large event-related slow potential (ERSP), but leaves the predominant 40- to 80-Hz frequencies unaltered. The ERSP was recorded with DC amplifiers (Beckman type R) from platinum-black electrodes, each of which was spaced at 0.5-mm intervals along the dorsal longitudinal surface and referenced to an electrode in sagittal bone.

In all cases the animal was *thoroughly* familiarized with the laboratory, personnel, and procedures before data were collected. Data were obtained and edited into 500-ms epochs for the 3 s before and after the beginning of the first inspiration of a faint novel odor (butanol); such an inspiration is detectable because it invariably has a shorter inspiratory interval compared to those observed during the previous 3-s control period. Respiration was monitored with a thoracic bellows and strain gauge.

The animal was accustomed to the delivery of several other faint odors (wintergreen and clove oil) for 2 weeks preceding the collection. An habituated odor (wintergreen) was presented 9 times during each of 10 days preceding the experiment. This habituated odor was presented for six trials during the same day preceding the novel stimulus (butanol). The intertrial intervals ranged from 3 to 10 min. The same experiment was repeated for 3 additional days to make possible observation of habituation to the butanol. A second novel odor (benzaldehyde) was used in a similar manner to that of butanol for observation of within-subject consistency.

The algorithm of Grassberger and Procaccia (1983) was used to analyze the data. The estimate of the the saturated correlation dimension (Hausdorff-Besicovitch number, HN) was calculated for each of the chosen data epochs. In most cases, 15 embedding dimensions were used. In one rabbit the simultaneously recorded epochs from 32 electrodes were each analyzed using this maximum number of embedding dimensions. We calculated the known HN of the Koch curve (log 4/log 3; Mandelbrot 1983) to test our use of the algorithm; the data were the apex and base points of all triangles generated, using three to eight iterations. Analysis of *white noise* epochs filtered to the bandpass (20–200 Hz), power spectrum (50–120 Hz), or principal components (50–80 Hz) of the bulbar data all had HNs that were clearly distinguished from those of the actual data passed through the same filter.

Each 500-ms epoch recorded from the bulb consisted of 320 digitized voltages produced by an A/D converter running at 500 kHz, but sampling at 640 Hz. Stationarity was presumed for each epoch. Although brief compared to epochs used by others, the 500-ms interval was chosen to increase the likelihood that stationarity would be maintained. Additionally, this choice seemed a reasonable interval because it takes approximately 500 ms for the cortex to process information, as demonstrated by (a) the duration of the short-latency event-related potentials (Haider et al. 1981) and (b) the minimum interval for neocortex to process low intensity input (Libet 1985).

Each of the voltage vectors in the epoch was subtracted from the value of all the rest to provide a set of absolute differences for dimensional analysis. The number of absolute differences, N accumulated within a range, R (of a given embedding dimension), is related to the correlation dimension, Dc, by the expression, $N = R^{Dc}$. If a linear relationship is detected in a plot of $\log N$ vs $\log R$, then Dc is quantifiable as the slope. The effects on slope produced by large and small R values (i.e., finite sets and near-neighbor correlations, Theiler 1986) were eliminated by visually determined cut-offs. When the embedding dimension is increased, the slope representing Dc will eventually saturate; that is, if an attractor exists. This saturated value of the slope is the estimate of the HN.

The tau interval is used to define the coordinates for the vectors for each embedding dimension. We selected tau by two methods: (a) the interval of the first zero-crossing of the autocorrelation function of the surface potential for each specific 500-ms epoch, or (b) one-half the mean zero-crossing interval for each epoch. For the data analyzed in this study, the two methods produced almost identical tau values, ranging from to 3.6 to 4.2 ms. No attempt was made to study the empirical dependency between tau and the saturated value of the HN. Convergence was clearly demonstrated in our data; the tau intervals used were small compared to the epoch lengths.

3 Results

Figure 1 A shows a 100-ms sample epoch (time) recorded simultaneously from 32 electrodes (space) during a preodor control condition. Four rows of eight electrodes each are shown. Note that during the control condition smooth voltage gradients are present, in both space and time (voltage is the upward coordinate). Figure 1 B shows the activities of the same electrodes 500 ms after inspiration of a novel odor. Note that the amplitude variations in time and space are no longer smooth. For example, the fusiform shape of the temporal signal seen in the leftmost trace in Fig. 1 A is not apparent in 1 B, and the peaks of the activities are often spatially isoelectric in 1 B, but decline smoothly, from row to row, in 1 A.

Figures 2–4 show the extensive analyses of data from a single animal. The traces in Fig. 2 show (from top to bottom): (a) the respiratory waveform, (b) the sniffing behavior (instantaneous change in respiratory rate), (c) the analog signal from a single sample electrode, and (d) the HNs for all of the 500-ms epochs for this single electrode. These data span the period just before and after the novel odor. Figure 3 shows the spatial distribution of the HNs calculated from the signals from 32 simultaneously recorded electrodes. These spatial data cover the same odor event as the temporal data shown in the previous figure. Note the stability of the HNs in the temporal domain in Fig. 2 and the homogeneity of calculated values in the spatial domain in Fig. 3. Only the electrodes in the upper left corner of Fig. 3 show any spatial heterogeneity. Postmortem examination revealed that these electrodes extended beyond the bulb and were not in contact with its surface, as were all of the rest in the array.

Figure 4 is similar to Fig. 2, but shows the HNs when the same animal has sniffed the butanol for the third time. In both cases the odor was injected into the

CONTROL　　　　　　　　　　　　NOVEL ODOR

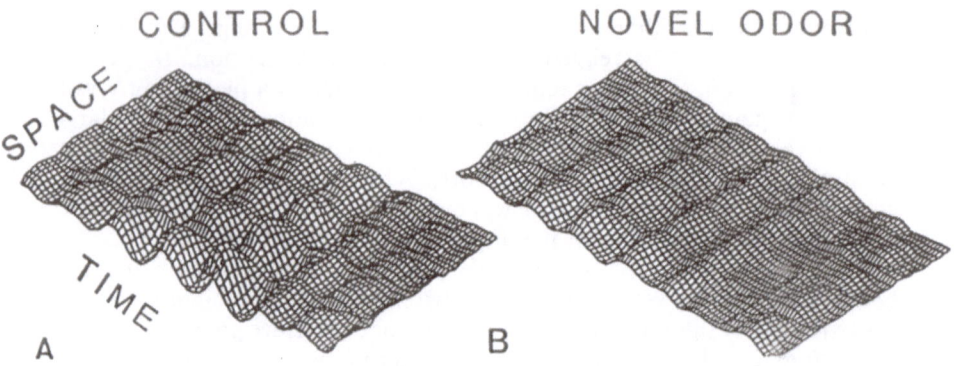

A　　　　　　　　　　　　　　　　　　B

Fig. 1 A, B. Simultaneously recorded data from 32 electrodes in an 8×8 array on the lateral surface of the right olfactory bulb. Four adjacent rows of eight electrodes each are observed in both space and time. The upward deflection is negative potential and the maximum peak-to-peak voltage shown is 325 µV. (A) A representative 100-ms epoch approximately 500 ms before the beginning of the first inspiration of a faint novel odor (butanol). (B) For the same electrodes, a 100-ms epoch approximately 1 s later

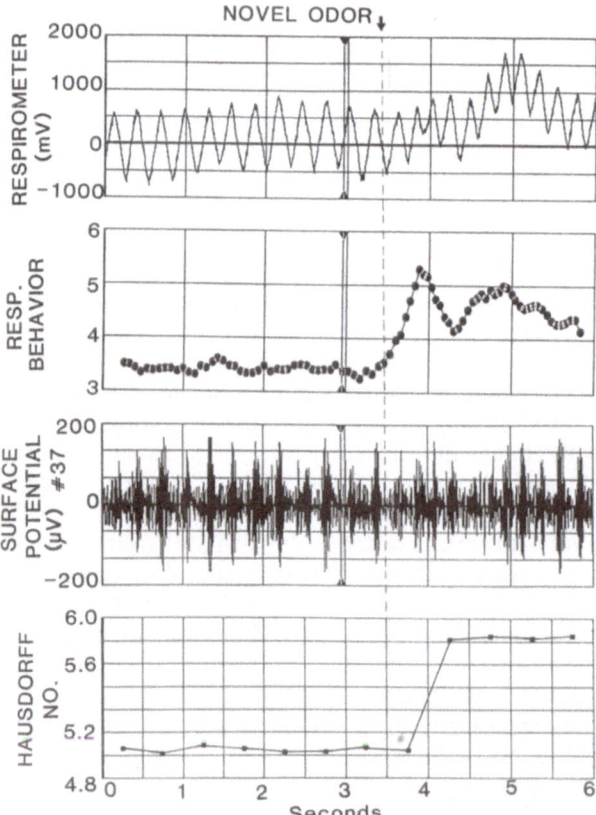

Fig. 2. Event-related alterations in respiration (*Respirometer*), instantaneous respiratory rate (*Resp. Behavior*), bulbar electric activity (*Surface Potential, position 37*), and surface potential dimensionality (*Hausdorff No.*). The perturbations were evoked together, in a conscious rabbit, by the first presentation of a novel odor (butanol). The algorithm used to analyze the respiratory signal is sensitive to instantaneous changes in slope and the units are respirations per second. The HNs are estimated for each of the 500-ms epochs indicated on the time scale, and each calculation was based on 15 embedding dimensions

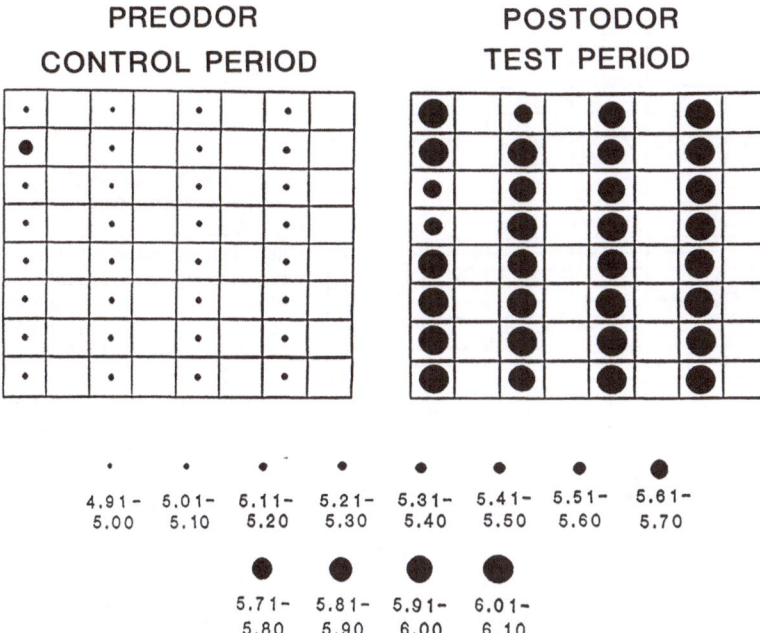

Fig. 3. Spatial distribution of HNs calculated for the surface potentials from 32 simultaneously recorded electrodes on the lateral surface of the olfactory bulb of the conscious rabbit, before and after the first presentation of a faint novel odor (butanol). The scale at the bottom identifies the interval (0.1 units of correlation dimension) in which each HN was embedded. Each preodor epoch was 500 ms in duration and began 1.0 s before the beginning of the first inspiration of the novel odor; each postodor epoch was 500 ms in duration and began 1.0 s after the beginning of the first inspiration of the novel odor

airstream of the nose cone at 3.0 s. Note in Fig. 4 that the behavioral reaction (Resp. Behavior) does not occur until the epoch beginning at 4.0 s. When the odor was novel both the behavior and the HN shifted together during the epoch beginning at 3.50 s, as seen in Fig. 2. In Fig. 4, however, there is a "droop" in the HN trace that begins at 3.50 s and anticipates the behavioral reaction.

Figure 5 A shows a sample plot of $\log R$ vs $\log N$ for all of the embedding dimensions used for one of the 500-ms epochs. Figure 5 B shows the plot of the slopes (i.e., the correlation dimension) vs the embedding dimensions. The linear portion of the slopes was determined visually, as is generally done by other investigators (Babloyantz and Destexhe 1986; Röschke and Başar, this volume; Graf and Elbert, this volume). Note that the linear portion of each slope and the saturation level of the group are both easily distinguished. Precision in determining the saturated value has been presumed, as the embedding dimensions number more than twice the value of the estimated HN.

Our version of the Grassberger and Procaccia algorithm was tested on several continuous functions, the dimensionalities of which are known. When used on Gaussian noise, no saturated value of the correlation dimension was observed within 15 embedding dimensions. The theoretical trajectory, using random

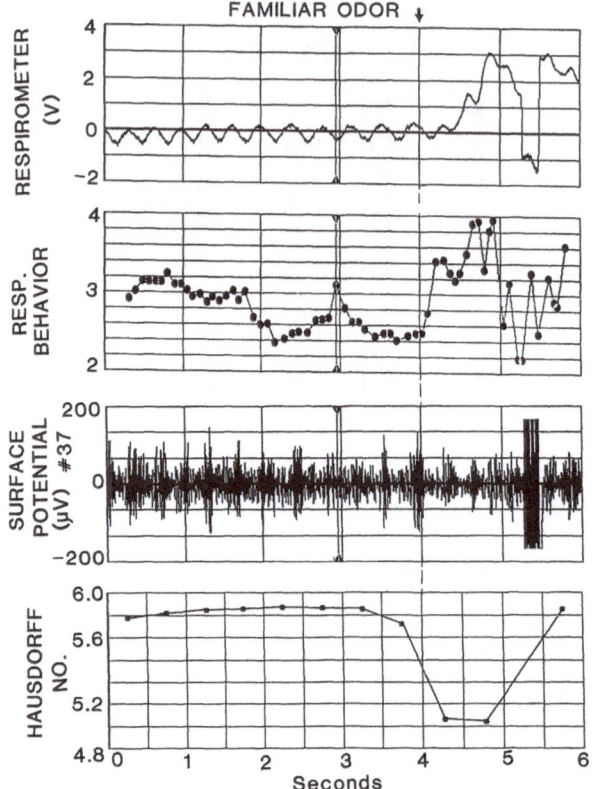

Fig. 4. Respiratory signal (*Respirometer*) and bulbar surface potential from a single electrode (position 37) and their analyzed perturbations (*Resp. Behavior* and *Hausdorff No.*) evoked in a conscious rabbit by the third presentation of an *habituated* odor (butanol). The data are from the same animal shown in Fig. 2. Note the difference in scale of the upper two traces. The HNs are estimated for each of the 500-ms epochs indicated on the time scale, and each calculation was based on 15 embedding dimensions; the epoch between 5.0 and 5.5 s was not analyzed because of the recording artifact seen in the surface potential record, caused by the animal shaking its head

numbers, is indicated in Fig. 5 B (NOISE). When used to analyze a sine wave, the saturated level occurred at HN = 1.000. When used to estimate the HN of the Koch curve (Fig. 6 C), the calculated value was very close to the theoretical value (i.e., 0.1265E + 01 was observed and 0.1262E + 01 is the theoretical value). It was determined empirically that at least 250 data points were needed for a close match between calculated and theoretical values, and that, for the Koch curve, increasing the number above 257 (four iterations) produced little additional improvement.

Table 1 presents the statistical evaluation for the changes in dimensionality evoked by a novel odor (butanol). A similar effect is observed, in the same animal, for the alternate novel odor (benzaldehyde), as shown in Table 2. Also seen in Table 2 is the effect of experience (habituation) on the direction of the dimensionality shift evoked by butanol. Note the small amount of variance detected, both temporally and spatially, for the control and odor conditions. Table 3 shows that between-subject variance is similar to that within-subjects.

A novel odor also produces the ERSP, a response which accompanies the spatial pattern changes on the surface of the olfactory bulb. Figure 7 shows an example of this response, which was evoked by butanol as the novel odor. Note that in all previous figures the bandpass filtered out this giant potential.

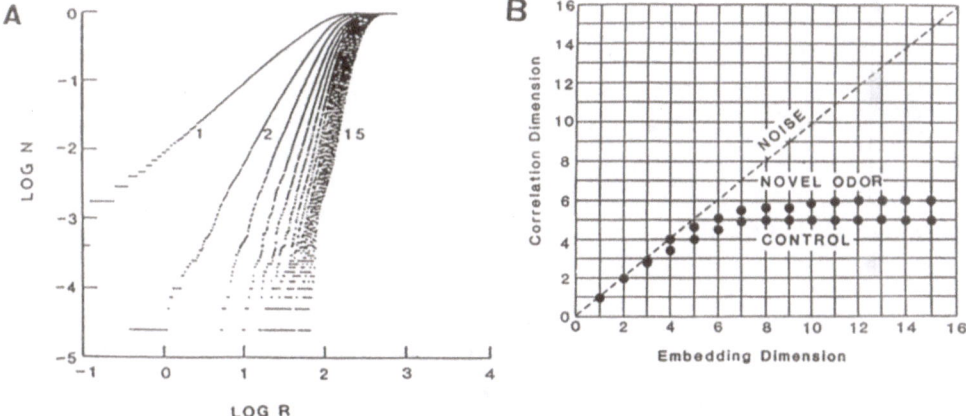

Fig. 5. (A) Sample data from a postodor epoch (first presentation of butanol) showing the accumulation of data points (N) as the range of the embedding dimension (R) is incremented within the set of difference scores. The latter were differences between all pairs of digitized voltages that occurred within the 500-ms epoch. The embedding dimension is indicated for the 1st, 2nd, and 15th traces. Units of R are the largest digitized value/1000 (approximately in units of 100 mV) and units of N are the accumulated number divided by the total number of data points. (B) Plot of the correlation dimension (i.e., slopes of the linear portion of each trace seen to the left) vs the embedding dimension. The preodor control data are also shown. The line labeled *NOISE* indicates the curve that would occur if the analysis had been made of a signal containing Gaussian noise. The asymptotic or saturated value indicates the HN as a point on the correlation dimension, which is a continuous scale. The mean and standard deviation of the last five points is our calculated estimate of the HN

Table 1. Hausdorff numbers[a] computed for 32, simultaneously recorded, 500-ms epochs during the control (C) and novel (N) stimulus conditions (EP, electrode position in the 8×8 array)

EP	C[b]	N[c]	EP	C	N	EP	C	N	EP	C	N
01[d]	5.03	6.02	17[d]	4.99	5.65	33	5.01	5.95	49	5.03	6.05
02[d]	5.63	6.04	18	5.03	6.02	34	5.01	5.98	50	5.11	6.05
03[d]	5.02	5.70	19	5.00	6.00	35	5.02	6.00	51	5.08	6.03
04[d]	5.03	5.07	20	5.03	6.03	36	5.02	6.06	52	5.03	6.03
05	5.05	6.01	21	5.04	6.01	37	5.03	6.03	53	5.07	6.01
06	5.05	6.04	22	5.04	6.05	38	5.05	6.04	54	5.03	6.06
07	5.05	6.06	23	5.05	6.04	39	5.05	6.04	55	5.07	6.06
08	5.05	6.06	24	5.04	6.00	40	5.06	6.06	56	5.05	6.04

[a] Numbers were calculated by the Grassberger and Procaccia (1983) algorithm; this algorithm was also used to calculate the dimensionality of the Koch curve. The latter is generated by a fractal function with a known HN, which is log(4)/log(3) or 1.265. The calculated value was found to be, within 5 digits of precision, the same as the known value.
[b] The 27 C values, with the five electrodes indicated by d excluded, have a mean and standard deviation of 5.04 ± 0.02, which is statistically significantly different from 5.00 ± 0.02 ($P < 0.01$).
[c] The 27 N values, with the five electrodes indicated by d excluded, have a mean and standard deviation of 6.03 ± 0.02, which is statistically significantly different from 5.04 ± 0.02 ($P < 0.01$, with or without the electrodes indicated by d included).
[d] These electrodes were found not to be in contact with the surface of the olfactory bulb.

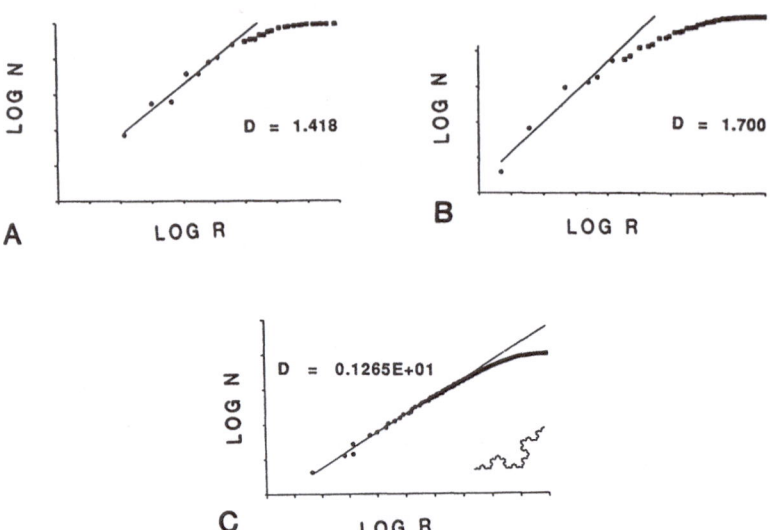

Fig. 6. Slope (*D*) of the plot of log *N* vs log *R* for the zero-crossing intervals in the 500-ms epochs just before (A) and 500 ms after (B) a novel odor (butanol). The slopes shown are for the first embedding dimension after saturation (embedding dimension of 2). The magnitude of the event-related increase in dimensionality for this animal is larger than that for the first rabbit. The same algorithm was used to calculate the slope for the data from the Koch curve function (C, 257 data points). For the latter, note that *D* is very close to the theoretical value, log 4/log 3, which is 0.1262E+01

Table 2. Effects of experience, within the same rabbit, on the direction of the event-related shift in the HN in response to the same and to different odor stimuli

Odor	Electrode position in array	HNs for control epochs[a]		HNs for test epochs[b]	
		C1	C2	T1	T2
The odors are novel (1st trial)					
Butanol	35	5.07	5.04	6.01	6.04
	37	5.03	5.03	6.02	6.04*
Benzaldehyde	35	6.06	6.04	7.08	7.08
	37	6.06	6.05	7.06	6.99*
The odor is familiar (3rd trial)					
Butanol	35	6.04	6.04	5.05	5.04
	37	6.07	6.05	5.06	5.04*

[a] C1 = 500-ms epoch 1.5 s before inspiration of odor. C2 = 500-ms epoch 1.0 s before inspiration of odor.
[b] T1 = 500-ms epoch 0.5 s after inspiration of odor. T2 = 500-ms epoch 1.0 s after inspiration of odor.
* $P < 0.01$, for all C and T comparisons during each odor.

Table 3. Within- and between-subject effects of a novel odor on the HNs associated with the pre- and postodor data (to reduce processing time, the dimensionality was calculated for the zero-crossing intervals of the analog signal, not its digitized amplitudes)

	Novel #1[a]		Novel #2[b]	
	HN, control	HN, test	HN, control	HN, test
Animal 1	0.90±0.004	0.95±0.002*	1.03±0.01	1.06±0.01*
Animal 2	1.41±0.004	1.70±0.004[c]*		

[a] Novel #1 = first presentation of butanol; HN, control and HN, test = brief epochs just before and after the inspiration of the novel odor.
[b] Novel #2 = first presentation of benzaldehyde.
[c] The estimate of the standard deviation was presumed to be the same as that of animal 1, as no suitable variance estimate was available for this rabbit.
* $P < 0.01$ (between control and test data from the same rabbit).

EVENT–RELATED

SLOW POTENTIAL

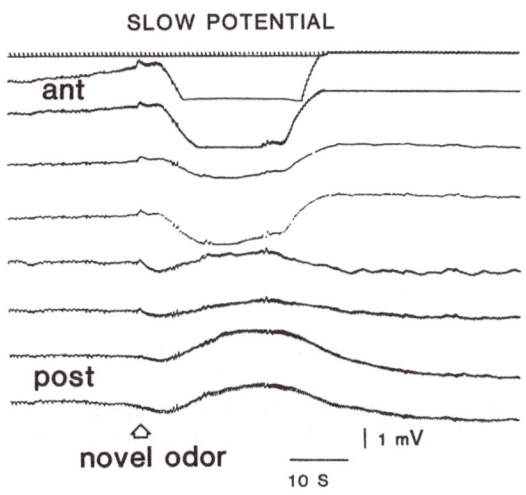

ant

post

novel odor

| 1 mV

10 S

Fig. 7. Effects of a faint novel odor (butanol) on the direct-coupled surface potentials of the olfactory bulb of the conscious rabbit. Note the large amplitude of the ERSPs. The recordings were made by DC-amplifiers from monopolar platinum-black electrodes (referenced to sagittal bone) distributed in an anterior–posterior (*ant, post*) line across the dorsolateral surface of the bulb; interelectrode spacing was at 0.5-mm intervals. Negative polarity is upwards

4   Discussion

Our current data, obtained from a simple-system model of cortical tissue, confirm previous observations (Babloyantz and Destexhe 1986, 1987; Röschke and Başar, this volume; Graf and Elbert, this volume) that a low-dimensional chaotic attractor can be constructed from a time-varying surface potential. These previous studies also showed that reduced cortical functioning, produced by (a) closing the eyes, (b) going to sleep, or (c) having an epileptic seizure, will result in the attractor having a reduced dimensionality. We showed that presenting a familiar odor to an awake alert rabbit also leads to a reduced dimensionality.

The unique feature of our data is that they show that the same odor can also evoke an *increase* in the dimensionality of the attractor, if it is novel rather than familiar. The direction of the evoked shift by the specific odor depends upon the context in which it is presented. That is, the dimensionality shifts of the attractor are *event related* and change in a direction related to the *learned* significance of the stimulus and not to its inherent attributes.

Our data show another feature not previously demonstrated for biological tissue: the same HN is observed *globally*, at all simultaneously recorded electrode sites. Even though the analog signals are all different at each site (Fig. 1), the HNs are all the same (Fig. 3). This spatial homogeneity appears to be the case generally, during all of the behavioral states, whether control or event related. It would therefore seem that a single, global, chaotic attractor exists everywhere throughout the olfactory bulb and that the dynamical process shifts to a new attractor (i.e., the HN changes) in relation to the odor event (novel vs habituated) and not to the specific odor itself.

The stability of the HNs in both the spatial and the temporal domains cannot be explained by the fact that they are all recorded monopolarly from a single reference electrode. Previous experience has shown that sagittal bone is a very good ungrounded referent; it manifests no potential above the amplifier noise level (Skinner and Yingling 1977).

5 Theoretical Considerations

5.1 Higher Cognitive Information Processing: Event-Related Shifting Among Chaotic Attractors

It is generally the event-related dimensionality increase or decrease that seems to be the focus for current neurobiological applications of mathematical dynamics to "higher cognitive functioning." The fact that the HNs generally appear to be fractions (i.e., have fractal dimensionalities) must also be of great importance. Why should HNs be both fractal and event related?

A reductionistic explanation for the fractal activity in the olfactory bulb dictates that the underlying molecular processes are themselves fractal and chaotic. This view is confirmed by the findings that physicochemical reactions (Bak 1985, 1986), enzymatic processes (Markus et al. 1985; Hess and Markus 1987), single ion-channel kinetics (Liebovitch et al. 1987; Liebovitch and Sullivan 1987), and membrane potentials (Chay and Rinzel 1985) all manifest behaviors characteristic of chaotic attractors. Furthermore, Liebovitch et al. (1987) state that their fractal model of ion-channel data is superior to previous stochastic models because it has fewer adjustable parameters, is more consistent with the dynamics of protein conformations, and fits better with their single-channel recordings from corneal endothelium and hippocampal neurons.

These nuts-and-bolts explanations, however, do not tell us *why* the fractal processes occur in cognitive mechanisms – was this selection by nature merely adventitious or was there a strategy involved? We propose that the reason for the existence of fractal dimensionalities of processes in cognitive systems is that only this type of dynamical process enables *data compression* before information storage. This hypothesis is supported by the realization that two coordinates given to the fractal function generating, say, the Mandelbrot set, can be deterministically projected into ten megabytes of exact pixel information. The problem is, how to do the *inverse;* that is, to compress the enormous amount of pixel information into the two coordinates and its underlying chaotic dynamical system.

Barnsley et al. (1987) have shown that geometric affine mapping applied to graphics data from "natural" objects (e.g., data depicting fern leaves or steam clouds) results in the determination of several different templates for the single data set. These abstracted templates of inherently fractal data can then be stored or transmitted at reduced bit rates and later used to reconstitute, deterministically, the original data set. The data-compression ratio can be as high as $10^6 : 1$. We propose that this property (i.e., the "inverse" process mentioned above) may explain why nature selected fractal dynamical systems in biological tissues to process cognitive information.

A stimulus (e.g., an odorant) has fixed physical attributes, but its accompanying context (e.g., the trial number) is variable each time the stimulus is presented to the subject. When the stimulus-plus-context is novel, more information must be stored than when the stimulus exists within a familiar context. This is because the thalamic gating mechanism *lets more information through to the cortex* when the stimulus is novel than when it is familiar (Skinner and Yingling 1977). Once the stimulus is either completely conditioned or habituated, evidence shows, inhibitory thalamic gating suppresses the ascent of irrelevant sensory information (Skinner and Yingling 1977). We now propose that when the stimulus is novel a higher-dimensional chaotic attractor is required to store the greater amount of information. This hypothesis is based on our present observations.

The affine mapping and attractor reconstitution mechanics described and demonstrated by Barnsley and Sloan (1988) require a relatively massive parallel processor to handle the fractal data. Cortical tissues, including the olfactory bulb, are well known massive parallel processors (Shepherd 1970). We propose that parallel processing in cortical tissues is not present for "backup" in case some circuits are damaged, as is often suggested, but rather is present because of the need for rapid on-line data compression.

Massive parallel processing by highly interconnected elements (i.e., "connectionist models") have emergent properties that resemble real behavioral learning (Hopfield and Tank 1986; Sejnowski and Rosenberg 1987). Skarda and Freeman (1987) suggested that fractal dynamics occur in the olfactory bulb, the structure of which conforms to the connectionist models. They claim, however, that features not attributable to the connectionist models are present in their data. They propose that making the connectionist models operate as chaotic dynamical systems may yield even more realistic emergent properties.

What is the next step, now that we know the olfactory bulb is a connectionist and chaotic dynamical system? We propose that biological empiricism may now afford the most rapid breakthrough in our understanding of the properties of this simple system, as it learns and remembers. Now that we have the idea that we are dealing with a single chaotic attractor whose dimensionality is event related, we may be able to generate a number of new hypotheses that will give us new insights. For example:

1. Does beta-receptor antagonism prevent the event-related attractor shifting?
2. Does a spatially fixed HN now allow us to "map" other regions of the attractor network, lying perhaps outside the bulb, each of which is also participating in the process?
3. If data compression is the raison d'être for a chaotic dynamical process, and if information storage is by a "compressed representation" and not a specific "engram," will we finally be able to recognize biologically stored information?

We seem to have come full circle to the neuroscience of the 1960s, but things are different; now we are armed with a theoretical perspective that incorporates a recently recognized strategy of nature.

5.2   The Brain–Heart Connection: A Common Pharmacology and Physiology?

Studies in comparative animal physiology suggest that the higher centers of the brain became a focus for natural selection and evolved to enable an organism to attend to its environment and simultaneously to prepare advance autonomic support for life-saving behaviors (Cannon 1931). Our laboratory has shown that the frontal lobes may be such an orchestrator of the sensory and autonomic channels (Skinner 1971; Skinner and Yingling 1977; Yingling and Skinner 1977). Bilateral output pathways from the frontal lobes to the thalamus selectively gate sensory input during behavioral attention. Separate descending projections to the brain stem initiate autonomic reactions in anticipation of the behavior they support. Of medical importance are our observations that the neural activities in the frontocortical–brain stem pathways are both necessary and sufficient to trigger lethal cardiac arrhythmias (Skinner and Reed 1981) and to maintain blood pressure elevations in experimental hypertension (Szilagyi et al. 1987).

Many electrochemical activities in the frontal lobes are event related (Skinner 1971; Skinner and Yingling 1976, 1977; Yingling and Skinner 1977). These include: extracellular slow potential shifts, norepinephrine release, cyclic AMP accumulation, potassium inactivation, and membrane depolarization (Skinner 1984; Skinner et al. 1983). Beta-receptor antagonists, which are cardiovascular drugs, reduce the magnitudes of all of these cortical responses; these same drugs also normalize blood pressure elevations and/or prevent lethal cardiac arrhythmias in human patients and animal models (Skinner 1971; Skinner and Yingling 1977; Yingling and Skinner 1977)

Long-term potentiation (LTP) is currently thought to be a useful model for synaptic learning and memory (Anderson and Wigstrom 1980; Swanson et al. 1982). Different cardiac drugs with separate modes of action (beta-receptor antagonism, catecholamine depletion) *disable* the development of LTP (Hopkins and Johnston 1984; Bliss et al. 1983). Our laboratory (Krontiris-Litowitz et al. 1985) has shown that yet another cardiac drug, with yet another mode of action (muscarinic agonism), also has an anti-LTP effect. Why both brain (LTP) and heart (arrhythmogenesis) processes are blocked in parallel by a variety of cardiac drugs constitutes a fascinating puzzle.

We have just demonstrated that novel odors evoke in bulbar surface potentials a dimensionality shift to a chaotic attractor that exists globally throughout the olfactory bulb. This result means that the spatial pattern of activities will be different between the control and postevent conditions. Our laboratory has shown that intrabulbar infusion of the cardiac antiarrhythmic drug, propranolol, will prevent the spatial pattern change that normally occurs between the control and postodor (novel odor) epochs (Gray et al. 1986). This latter observation is evidence that propranolol also blocks event-related dimensionality shifts. Could an arrhythmia be the result of a sudden shift in a global cardiac attractor?

The cellular and subcellular oscillators that produce the cardiac action potential are known to be characterized by a chaotic attractor (Guevara et al. 1981). The nonoscillatory cardiac tissues also manifest nonlinear dynamics (Chialvo and Jalife 1987). Goldberger et al. (1984) suggested that cardiac arrhythmias could arise because of a change in a *parameter* of the chaotic attractor in the heart. We propose that the attractor itself changes, i.e., its dimensionality shifts.

Based on the nonretraction theorem in topology, Winfree (1983, 1987) suggested that the orthogonal overlap of voltage and phase gradients in myocardial tissue will result in at least one point, called a phase singularity, which is chaotic; i.e., chaotic in the sense that application of a suitably graded stimulus in this neighborhood puts the tissue in a state from which it cannot spontaneously recover to normal beating at a predictable time. He showed with his model that this singularity is surrounded by rotating waves of isochrons (loci of equal activation time); these waves in fact have been observed in cardiac tissue during arrhythmogenesis. He further demonstrated that the relative size of the phase singularity can be increased. This increase, he suggested, could lead to an increase in the vulnerability of the heart of arrhythmogenesis.

Three observations from our laboratory demonstrate ways to increase cardiac vulnerability to arrhythmogenesis. These observations may suggest ways to increase the relative size of Winfree's phase singularity.

1. Descending activity in cardiac nerves is necessary for the occurrence of the lethal arrhythmia, ventricular fibrillation (Skinner and Reed 1981). Winfree's model suggests that a voltage gradient, probably produced by nerves and orthogonal to the isochron gradient, is necessary to produce the phase singularity; without current being injected into the heart by the nerves, the singularity could not exist.
2. Stimulation of sympathetic and parasympathetic activity combined produces the most malignant type of descending neural activity (Skinner et al. 1983); such dual activation of the heart certainly would change the shape of the phase singularity and it could be expected to increase its relative size.
3. The event-related increase in the dimensionality of an attractor in the brain (represented by the present data) may be projected to the heart (e.g., through the frontocortical–brain stem pathway), in which case the relative size of the phase singularity may be increased – i.e., dimensionality increase may be related to phase singularity increase.

6   General Conclusions

All of the olfactory bulb appears to be used all of the time for the processing of odor signals. This result is an electrophysiological example of Lashley's well-known principle of mass action. The main problem in studying the neurophysiology of mass action is that we do not yet know how to deal with the enormous number and complexity of interactions that occur among neurons during a "higher cognitive process." Nonlinear mathematical dynamics may be the tool we need to study such intense interactive processes. When we gain a clearer understanding of the biological significance of fractal processes and dimensionality shifts, comparable to our understanding of some physical systems (Gleick 1987), then we may come to a completely different (perhaps even revolutionary) view of the basic cognitive process. The same process may operate in the heart as well as in the brain, and, if understood, could lead to new insights and interventions for dealing with the problems of cardiovascular malfunction.

7   Summary

After familiarization with the recording chamber, each of two rabbits was presented with a novel odor while recording from an 8×8 electrode array on the surface of the olfactory bulb (bandpass: 20–200 Hz, 3 dB). The Hausdorff number (HN) is a mathematical descriptor of the number of independent dimensions that appear to generate an observed dynamical process; it can be fractional. Preodor epochs (500 ms each) had HNs that were the *same*, both spatially (mean HN = 5.04 ± 0.02 #SD) and temporally (mean HN = 5.04 ± 0.02). After the odor, the HNs briefly increased ($P < 0.01$) to new values (6.03 ± 0.02), lasting for some seconds, before returning to control. Following only two trials of habituation, there was a *decrease* in the odor-evoked HNs. All HNs observed were fractional, with statistical significance. All electrodes showed the same HNs at all times. All HNs had clearly saturated correlation dimensions (using 15 embedding dimensions); this convergence occurred even though only 320 digitized data points were in the epoch. These results suggest that the electric activities throughout the bulb may be characterized spatially by a single global chaotic attractor, the dimensionality of which is event related.

References

Anderson P, Wigstrom H (1980) Possible mechanisms for long-lasting potentiation of hippocampal synaptic transmission. In: Tsuko Y, Aghanoff BW (eds) Neurobiological basis of learning and memory. Wiley, New York, pp 17–47
Babloyantz A, Destexhe A (1986) Low-dimensional chaos in an instance of epilepsy. Proc Natl Acad Sci USA 83:3513–3517

Babloyantz A, Destexhe A (1987) Strange attractors in the human cortex. In: Rensing L, an der Heiden U, Mackey MC (eds) Temporal disorder in human oscillatory systems. Springer, Berlin Heidelberg New York, pp 48–56 (Springer series in synergetics, vol 36)

Bak P (1985) Mode-locking and the transition to chaos in dissipative systems. Phys Scr 9:50–58

Bak P (1986) The devil's staircase. Phys Today 86:38–45

Barnsley MF, Sloan AD (1988) A better way to compress images. Byte 13:215–223

Barnsley MF, Massopust P, Strickland H, Sloan AD (1987) Fractal modeling of biological structures. NY Acad Sci 504:179–194

Bliss TVP, Goddard GV, Rives M (1983) Reduction of long-term potentiation in the dentate gyrus of the rat following selective depletion of monoamines. J Physiol (Lond) 334:475–491

Cannon WB (1931) Again the James-Lange and the thalamic theories of emotion. Psychol Rev 38:281–295

Chay TR, Rinzel J (1985) Bursting, beating, and chaos in an excitable membrane model. Biophys J 47:357–366

Chialvo DR, Jalife J (1987) Non-linear dynamics of cardiac excitation and impulse propagation. Nature 330:749–752

Farmer JD (1982) Dimension, fractal, measures, and chaotic dynamics. In: Haken H (ed) Evolution of order and chaos, Springer, Berlin Heidelberg New York

Freeman WJ, Schneider WS (1982) Changes in spatial patterns of rabbit olfactory EEG with conditioning to odors. Psychophysiology 19:44–56

Gleick J (1987) Chaos: making a new science. Penguin, New York

Goldberger AL, Shabetai R, Bhargava V, West BJ, Mandel AJ (1984) Nonlinear dynamics, electrical alternans, and pericardial tamponade. Am Heart J 107:1297–1299

Grassberger P, Procaccia I (1983) Measuring the strangeness of strange attractors. Physica [D]9:183–208

Gray CM, Freeman WJ, Skinner JE (1984) Associative changes in the spatial amplitude patterns of rabbit olfactory EEG are norepinephrine-dependent. Soc Neurosci Abstr 10:121

Gray CM, Freeman WJ, Skinner JE (1986) Chemical dependencies of learning in the rabbit olfactory bulb: acquisition of the transient spatial pattern change depends on norepinephrine. Behav Neurosci 100:585–596

Guevara MR, Glass L, Shrier A (1981) Phase locking, period-doubling bifurcations and irregular dynamics in periodically stimulated cardiac cells. Science 214:1350–1353

Haider M, Groll-Knapp E, Ganglberger JA (1981) Event related slow (DC) potentials in the human brain. Rev Physiol Biochem Pharmacol 88:125–197

Hess B, Markus M (1987) Order and chaos in biochemistry. Trends Biochem Sci 12:45–48

Hopfield JJ, Tank DW (1986) Computing with neural circuits: a model. Science 233:626–633

Hopkins WF, Johnston D (1984) Frequency-dependent noradrenergic modulation of long-term potentiation in the hippocampus. Science 226:350–352

Krontiris-Litowitz J, Skinner JE, Birnbaumer L (1985) A muscarinic agonist (Ethmozine) prevents long-term potentiation in the hippocampal slice. Soc Neurosci Abstr 11:781

Libet B (1985) Unconscious cerebral initiative and the role of conscious will in voluntary action. Behav Brain Sci 8:529–566

Liebovitch LS, Sullivan MJ (1987) Fractal analysis of a voltage-dependent potassium channel from cultured mouse hippocampal neurons. Biophys J 52:979–988

Liebovitch LS, Fischbarg J, Koniarek JP, Todorova I, Wang M (1987) Fractal model of ion-channel kinetics. Biochim Biophys Acta 896:173–180

Mandelbrot BB (1983) The fractal geometry of nature. Freeman, New York

Markus M, Kuschmitz E, Hess B (1985) Properties of strange attractors in yeast glycolysis. Biophys Chem 22:95–105

Sejnowski TJ, Rosenberg CR (1987) Parallel networks that learn to pronounce English text. Complex Syst 1:145–168

Shepherd GM (1970) The olfactory bulb as a simple cortical system: experimental analysis and functional implications. In: Schmitt FO (ed) The neurosciences: second study program. Rockefeller University Press, New York, pp 539–552

Skarda CA, Freeman WJ (1987) How brains make chaos in order to make sense of the world. Behav Brain Sci 10(2):161–173

Skinner JE (1971) Abolition of a conditioned, surface-negative, cortical potential during cryo-genic blockade of the nonspecific thalamo-cortical system. Electroencephalogr Clin Neuro-physiol 31:197–209

Skinner JE (1984) Central gating mechanisms that regulate event-related potentials and behav-ior. In: Elbert T, Rochstroh B, Lutzenberger W, Birnbaumer N (eds) Self-regulation of the brain and behavior. Springer, Berlin Heidelberg New York, pp 42–58

Skinner JE, Reed JC (1981) Blockade of a frontocortical-brainstem pathway prevents ventricular fibrillation of the ischemic heart in pigs. Am J Physiol 240:H1156–H1163

Skinner JE, Yingling CD (1976) Regulation of slow potential shifts in nucleus reticularis thalami by the mesencephalic reticular formation and the frontal cortex. Electroencephalogr Clin Neurophysiol 40:288–296

Skinner JE, Yingling CD (1977) Central gating mechanisms that regulate event-related potentials and behavior: a neural model for attention. In: Desmedt JE (ed) Progress in clinical neuro-physiology, vol 1. Kärger, Basel, pp 30–69

Skinner JE, Beder S, Entman ML (1983) Psychological stress activates phosphorylase in the heart of the conscious pig without increasing heart rate and blood pressure. Proc Natl Acad Sci USA 80:4513–4517

Swanson LW, Teyler TJ, Thompson RF (eds) (1982) Hippocampus long-term potentiation: mechanisms and implications for memory. Neurosci Res Program Bull 20:613–769

Szilagyi JE, Taylor AA, Skinner JE (1987) Cryoblockade of the ventromedial frontal cortex re-verses hypertension in the rat. Hypertension 9:576–581

Theiler J (1986) Spurious dimension from correlation algorithms applied to limited time-series data. Phys Rev [A]34:2427–2432

Winfree AT (1983) Sudden cardiac death: a problem in topology. Sci 248:144–161

Winfree AT (1987) When time breaks down: the three-dimensional dynamics of electrochemical waves and cardiac arrhythmias. Princeton University Press, Princeton

Yingling CD, Skinner JE (1977) Gating of thalamic input to cerebral cortex by nucleus reticularis thalami. In: Desmedt JE (ed) Progress in clinical neurophysiology, vol 1. Karger, Basel, pp 70–96

Dimensional Analysis of the Waking EEG*

K. E. Graf and T. Elbert

1 Introduction

The theory of nonlinear systems has become increasingly useful and relevant to the study of empirical dynamics. When a system produces irregularity in one or more of its variables, it is of interest whether this behavior results from *randomness* (meaning that the number of degrees of freedom is infinite) or whether a finite, and possibly small, number of degrees of freedom has produced the chaos (meaning that the system is deterministic). Our understanding of deterministic systems was greatly enhanced when Lorenz (1963) discovered that a simple system with as few as three differential equations can generate totally irregular fluctuations of the system's variables – a phenomenon nowadays generally referred to as *deterministic chaos*. The prominent features of chaos are *unpredictability* over extended time periods, and *sensitive dependence on initial conditions*. Once started with specific initial values, the system's future might be totally different from what it would have been if it had been started under slightly different initial conditions. Chaos may not be the ultimate description for a system's irregular dynamic. As outlined by Rössler (1983), more complex structures "beyond chaos" may await discovery.

The scientific areas in which analyses in terms of chaotic but deterministic behavior have been useful range from physics (e.g., the dynamics of fluids, Swinney and Gollub 1986), lasers (Albano et al. 1986), and chemistry (e.g., nonequilibrium reactions, Roux et al. 1983), to biology (the dynamics of population growth, Schaffer and Kot 1986) and physiology (e.g., certain kinds of altered breathing and cardiac rhythms, Glass et al. 1986; cellular metabolism, Rapp 1986). The suggestion that the activity of single neurons as well as cell assemblies may be described by means of nonlinear system analysis (Rapp et al. 1985) leads us to the dimensional analysis of the EEG.

The EEG originates primarily in cerebral cortex. It reflects summed electrical activity of billion of neurons. Their interconnections form the most complex system of which we are aware. It has been argued that the brain has so many degrees of freedom that we never will be able to construct appropriate models which might elucidate the workings of our mind. But consider a gas in a closed chamber in which innumerable molecular collisions are taking place all the time. Each particle travels along a different path. Nevertheless, there are a number of macroscopic relationships which are quite able to describe properties of such a system like pressure, temperature, and their relationship in the *general law for gases*. This law was discovered and described on a purely macroscopic level. On the other-

* Originally published in Başar E, Bullock TH (eds) Brain dynamics. Springer, Berlin Heidelberg New York, pp 174–191 (Springer series in brain dynamics, vol 2). Cross references refer to that volume.

hand, if we look at a small particle being pushed around in Brownian motion by molecules within the gas, we cannot detect the rules which govern such a process on a macroscopic basis alone, although the motion is visible on the macroscopic level.

Extracting parameters from the EEG corresponds to a macroscopic measurement. The attempts to set up rules among EEG patterns and brain wave components or between such measures and behavioral elements have not been particularly successful in the past. This has led pessimists to view the EEG like the noise of the brain motor, very much like the seemingly random Brownian motion. But even for optimists who untiringly try to uncover relations which might predict EEG behavior, the (so far primary) dimensional analysis of the EEG has exceeded their prior expectations. Babloyantz (Babloyantz et al. 1985; Babloyantz and Destexhe 1986) was the first to report low dimension values in the human EEG. She detected dimensions of 4.05 and 4.27 for slow wave sleep stage 4 and 4.99–5.03 for non-REM stage 2. Others (Mayer-Kress and Layne 1986; Albano et al. 1986; Dvorak and Siska 1986; Rapp et al. 1986) reported dimensions lower than 10 during the relaxed waking state with eyes either closed or open. If such analyses are confirmed, one could conclude that the EEG results from a "rather simple" deterministic system with only a couple of variables to be determined. Based on neuroscientific evidence, Elbert and Rockstroh (Elbert and Rockstroh 1987; Rockstroh and Elbert 1988; Elbert 1987) have outlined ways by which a chaotic time series of electrical activity might be generated by the brain's regulatory processes. They conclude that nonlinear system analysis of the EEG is worthy of further study.

The present paper represents such a study. After a brief methodological introduction we will compare two ways of evaluating the time lag needed to construct the "equivalent" state space of the EEG generating system, and then go on to estimations of the EEG dimensionality.

2   Methods I: Optimal Time Lag and Data Preparation

The *state space* is a useful concept for visualizing the system's dynamic behavior. It is an abstract space whose coordinates are the degrees of freedom of the considered system, i.e., the number of independent variables. We cannot, however, construct directly such a state or phase space for the part of the brain which generates the EEG, since we do not know its specific coordinates or its degrees of freedom. Fortunately, it is apparently possible to employ a technique to reconstruct an equivalent state space, since a single time series of one variable [like the EEG(t)] can be interpreted as containing a condensed image of the whole system which has generated the time series. [1]

---

[1] The EEG refers to a voltage which is measured between any two electrodes on the head as a function of time. All this electrical activity could arise from one single system (in the sense of system theory). Then the information, contained in EEG time series recorded from different locations would be largely redundant, but the EEG record with the highest amplitudes would provide most information. For this reason we chose vertex–ear recordings for the present analyses. On the other hand, multiple sources in different and rather independent systems of the brain may contribute to the EEG. Then differing electrode locations may pick up activity predominantly generated in different systems, although volume conduction will always cause some overlap in recorded activity, simulating again a single complex system, rather than an ensemble of different subsystems.

Given a single time series like EEG(t) from just one location, how is it possible to resurrect the complete dynamics of the generating system which possibly includes quite a number of independent variables? Takens (1981) and Packard et al. (1980) proposed reconstructing the state space by means of *time delays* τ: The values measured at fixed time delays EEG(t), EEG($t+\tau$), EEG($t+2\tau$) ... are treated as though they characterize new variables. The delayed values then define a single point in a multidimensional state space and thus a vector. Let EEG(t_1), EEG(t_2), EEG(t_3), ..., EEG(t_n) be the empirical scalar time series of the digitized EEG. The reconstructed vectorial time series will then be:

$$
\left\{
\begin{pmatrix}
\text{EEG}(t_1) \\
\text{EEG}(t_1+\tau) \\
\text{EEG}(t_1+2\tau) \\
\text{EEG}(t_1+3\tau) \\
\vdots \\
\text{EEG}(t_1+(d_e-1)\tau)
\end{pmatrix}
\begin{pmatrix}
\text{EEG}(t_2) \\
\text{EEG}(t_2+\tau) \\
\text{EEG}(t_2+2\tau) \\
\text{EEG}(t_2+3\tau) \\
\vdots \\
\text{EEG}(t_2+(d_e-1)\tau)
\end{pmatrix}
\begin{pmatrix}
\text{EEG}(t_3) \\
\text{EEG}(t_3+\tau) \\
\text{EEG}(t_3+2\tau) \\
\text{EEG}(t_3+3\tau) \\
\vdots \\
\text{EEG}(t_3+(d_e-1)\tau)
\end{pmatrix}
\cdots
\right\}
$$

with d_e as the *embedding dimension* to reconstruct the dynamics. Connecting the points subsequent in time creates a *trajectory* in the state space. The set of such trajectories for all of the possible initial conditions forms a phase portrait. If trajectories starting from different initial values are attracted to a discrete region within the state space with a lower dimension (i.e., a subspace), this region is called an *"attractor."* (An attractor may have a fractal dimension, and is then called "chaotic" or "strange"; see Crutchfield et al. 1986; Graf 1986, for an introduction to chaotic attractors.) The attractor characterizes the dynamically stable solution which the system will adopt in the long run. Hence the attractor is a manifestation of the system's self-organization. As shown by Whitney (1936), a complete embedding of the system's dynamics is ascertained for values of d_e greater than 1 plus 2 times the dimension of the dynamics, i.e., the attractor. In many cases, however, a smaller value of the embedding dimension d_e is sufficient. The lowest embedding dimension which is possible for a given attractor determines the number of degrees of freedom of the system which has generated the corresponding time series.

It is clear that the time delay τ plays a crucial role when reconstructing the attractor. For practical reasons, τ is chosen as a multiple of the sampling time Δt, which is the interval between two successive measurements. Takens (1981) has based his proof for the adequacy of the time delay technique on two assumptions: the time series is infinite, and the signal is not contaminated with noise. Thus the value of the time delay should have no effect on the outcome. These suppositions, however, can be hardly satisfied by empirical data. Therefore, only a certain range of time delays will adequately represent the dynamics of the system considered. More precisely, the assumption that successive measurements contain new information, irrespective of the length of the time intervals between them, is not valid, since precision of measurement is limited. If the time lag τ is too short, successive points in phase space will appear along a straight line, stretching the attractor artificially along the diagonal. Choosing too large a time lag τ, on the other hand, will make recognition of the attractor difficult, since it is a property of chaotic attractors that small errors in measurement blow up exponentially over time.

There are two possibilities for obtaining an estimate for a valid τ value from a measured time series. The first one is to set τ equal to the smallest time lag for which the *autocorrelation function* is zero or to the first local minimum if that is earlier than the zero point. In the latter case the local minimum will be positive. The second possibility is to choose the first local minimum of the *mutual information function*.

The concept of mutual information as developed by Fraser and Swinney (1986) evolved from information theory in an attempt to answer the question: Given the measured value of a function $f(t)$ at time t, how many bits of $f(t+\tau)$ can be predicted on the average? $f(t)$ and $f(t+\tau)$, the coordinates of the state space, will be most independent where the mutual predictability reaches a minimum. The calculation of mutual information basically works in terms of *simple* and *conditional entropy:*

Assume a signaling system S sends a series of messages. All the messages sent are drawn from a certain set of possible messages $s_1, s_2, s_3, \ldots, s_n$. By the frequency of each of the possible messages s_i, the associated probabilities $P_s(s_1), P_s(s_2), \ldots, P_s(s_n)$ can be determined. The average amount of information gained from a detection that specifies a message of S is called the *entropy* H of the signaling system S:

$$H(S) = - \sum_{i=1}^{n} P_s(s_i) \cdot \log P_s(s_i) .$$

If the base of the logarithm is 2, the units of $H(S)$ are bits. In other words, $H(S)$ specifies how many yes/no questions are needed in order to identify one given message, provided the questioning strategy is optimal. As a next step let us consider two coupled signaling systems $[S, Q]$. If the message of S is known, a prediction of the message of Q will be possible. The goal here is to quantify the precision of the prediction, so we must examine the conditional entropy $H(Q/S)$ first. Based on the detection of S, we will make a prognosis of Q, and $H(Q/S)$ will tell us the average increment for the information in bits, which we will obtain from a detection that specifies Q. According to Fraser and Swinney (1986) we can write

$$H(Q/S) = H(S, Q) - H(S) \quad \text{whereby}$$

$$H(S, Q) = - \sum_{i,j} P_{sq}(s_i, q_j) \cdot \log P_{sq}(s_i, q_j)$$

P_{sq} is the joint probability distribution of the coupled systems. Computing the difference between the simple entropy $H(Q)$ and the conditional entropy $H(Q/S)$, i.e., subtracting the residual information content of Q – given the knowledge of S – from the total information content of Q, we obtain the mutual information $MI(Q,S)$.

$$MI(Q, S) = H(Q) - H(Q/S)$$
$$= H(Q) + H(S) - H(S, Q) = MI(S, Q) .$$

Table 1. Subject characteristics

Sex	Age	Medication (per day)	Local minimum (τ in ms)	Results illustrated in
Epileptic patients				
m	17	4 Tegretal 4 Zentropil 2 Luminal	60	
m	43	2 Orphiril 3 Tegretal	70	
m	35	3 Tegretal 2 Zentropil	70	Figs. 1 (bottom) and 5
f	15	1.5 Phenhydan 3.5 Timonil Retard 5 Soxinutin	60	
m	16	2 Tegretal 3 Ergenyl	90	Fig. 6
m	25	1 Maliasin 1 Petinutin 1 Orphiril 1 Phenhydan	50	
f	15	None	30	Figs. 1 (bottom), 2, and 3
Healthy controls				
m	28	–	60	
m	44	–	50	Fig. 4

This value provides the answer to our question; namely, it quantifies how precisely on the average, the message of Q can be predicted from knowing the message of S. If the two systems are closely coupled, a precise prediction of the messages of one system will be possible from the messages of the other system. If the dependency among systems is low, their mutual predictability will be poor, and *MI* will be small. A detailed procedure for the calculation of *MI* has been provided by Fraser and Swinney (1986). We have applied this procedure, making the assignment $[S, Q] = [\text{EEG}(t), \text{EEG}(t+\tau)]$. [2]

The advantage of the mutual information over the autocorrelation function lies in the fact that the former considers all kinds of relations, not only the linear ones, as the latter does. [3]

---

[2] The procedure requires the length of the time series to be a power of 2. We have chosen $2^{13} = 8192$ data points.

[3] For embedding dimensions greater than 2 it is required to consider all the possible dependencies, now among a much larger number of coordinates, in order to find the optimal time delay. Therefore mutual information could be generalized the redundancy, which deals with any number of variables (Fraser 1986). But then a very high number of data points would be required, which is generally not available, i.e., such calculations are generally not possible. The solution most often accepted is to assume one and the same time lag τ for all of the embedding dimensions. Then τ can be calculated by using an embedding dimension of 2.

Both methods of estimating the optimal time lag τ were applied for 81.92-s segments of EEG selected to be free of movement or eye movement artifacts, but not blink artifacts. Our data were recorded from Cz vs. linked earlobes during the waking states of two healthy human subjects and six epileptic patients. Vertical eye movements were recorded from electrodes below and above the left eye. EEG and EOG were amplified by means of a Beckmann type R polygraph set for a band pass from 0.005-30 Hz (30 s time constant). All patients but one were under long-term treatment with antiepileptic medication (see Table 1). All were drug resistant, having from one to several seizures per week. After electrode hook up, subjects were told to relax for 5 min. The session started with another 5-min relaxation period, followed by 3 min of hyperventilation and a final 3-min rest period. Subjects were instructed to keep their eyes open and to fixate a 3×3 cm cross positioned 2 m in front of them. The EEG segments were selected from the initial 5-min relaxation period and for patients additionally during times when seizure-like activity appeared. EEG records were also obtained while the subjects were told to relax with their eyes closed. The EEG was sampled at a rate of 100/s. (The results for the conventional analysis of these data are described in more detail by von Bülow et al. 1988.)

3   Results for the Time Lag τ

The mutual information function was calculated for three 81.92-s long EEG segments, chosen from the beginning (I), the middle part (II), and the end (III) of the initial 5-min relaxation period. How stable is the first minimum during such a time period? Figure 1 (top) depicts the mutual information of an unmedicated patient's EEG, plotted against the time lag τ. For all the three periods, the minimum lies around 30–40 ms. This value is much shorter than the one of a medicated epileptic patient, whose mutual information function is also shown in Fig. 1 (bottom). His minimum is also stable; it lies, however, between 60 and 70 ms. For the six patients under medication, the values range from 50 to 90 ms (see Table 1). Despite the considerable variance between subjects, the minimum turned out to be completely stable within subjects, at least within the 10 ms resolution used. As indicated in Fig. 2, this stability also holds over a period of 5 weeks, at least for a given experimental condition.

For 22 EEG segments recorded during the waking state, we compared the two possible ways the time lag τ was selected: from the mutual information function or from the autocorrelation function. The results were consistently the same: mutual information will reach its first local minimum aproximately at that point where the autocorrelation intersects with the abscissa. If there is a first (positive) minimum of the autocorrelation function, as in Fig. 3, mutual information will have its minimum at this value of τ. For some cases, however, the mutual information function does not exhibit a pronounced local minimum during its slow decay with increasing time lag.

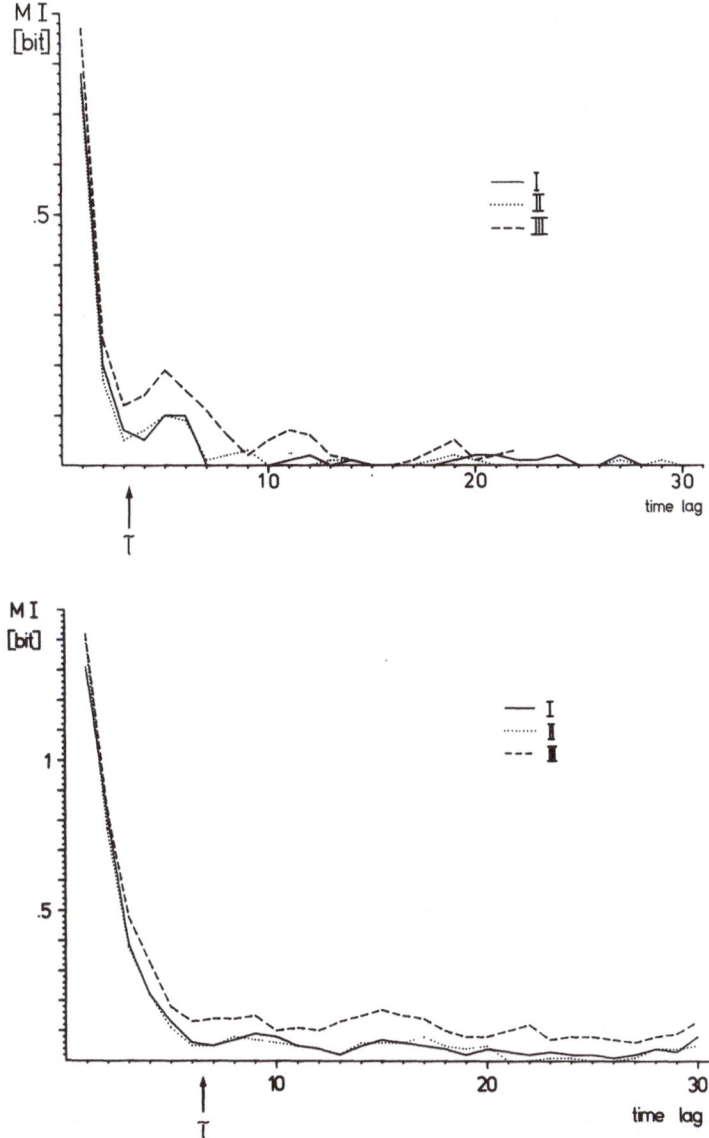

Fig. 1. The mutual information (MI) of the EEG is plotted as a function of the time lag τ. The units on the *abscissa* represent multiples of the 10 ms sampling unit. Results are illustrated for one unmedicated epileptic patient (*upper part*) and one under extensive pharmacotherapy (*lower part*). The three different lines in each plot represent EEG segments of one subject chosen from the beginning (I), middle part (II), and end (III) of a 5-min rest period. The length of the EEG segments is 81.92 s, corresponding to 8192 points. *Arrows* mark the optimal time lag τ, i.e., the first local minimum

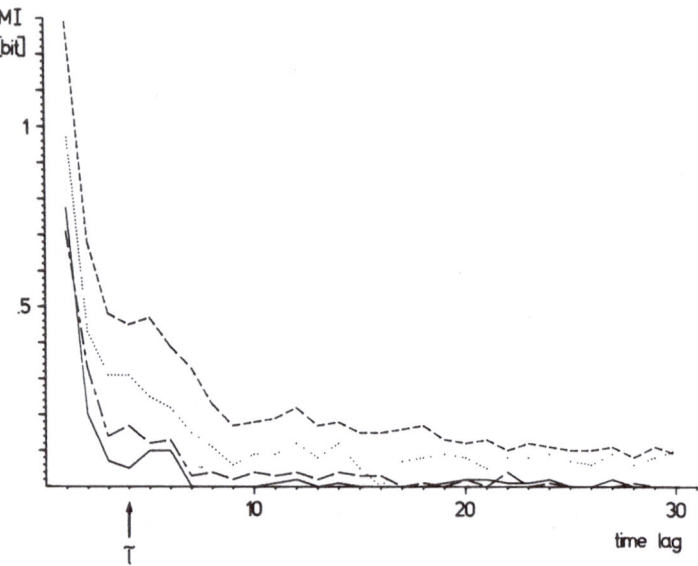

Fig. 2. Mutual information as a function of the time lag τ, calculated from four different EEG segments which were recorded from an unmedicated patient on four different occasions over a period of 5 weeks

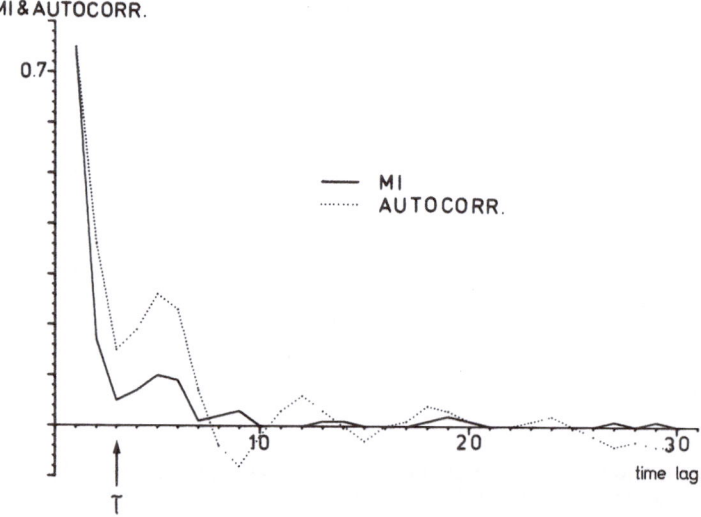

Fig. 3. Example of the mutual information function (units in bits) and the autocorrelation function (units correspond to correlation coefficients) computed for the same EEG segment. Both methods yield the same result for the optimal time lag τ

3.1 Discussion

The two criteria for choosing the optimal time lag τ yield equal results, at least during the waking state. Obviously, nonlinearity in the autocorrelations of the time series does not play a major role in the EEG analyses. This is not necessarily the case for other irregular time series as shown, for example, for the Roux attractor (by Fraser and Swinney 1986).

It seems sufficient to extract optimal EEG time delays from the EEG autocorrelation function, which also circumvents the problem of identifying τ when there is no distinct local minimum in the MI function.

As indicated in Table 1, the optimal time lag τ appears to be longer in patients under anticonvulsant medication. Although the present data base is too small for a definite conclusion, this result fits with observations of delayed event-related potentials, a generally slowed EEG, and a slower reaction time under anticonvulsants (Rockstroh et al. 1987, von Bülow et al. 1988). It is quite possible that a kind of cycle time, inherent in the regulation of brain activity (Elbert 1987; Elbert and Rockstroh 1987), is slowed down in subjects receiving antiepileptic medication. It is interesting that τ is shortest for the unmedicated epileptic patient, being below the values for the normal controls, whereas medication brings the values back towards the range for healthy controls. According to this, the patient with $\tau = 90$ ms must have been overmedicated. As a matter of fact, after obtaining the EEG result, analyses of the plasma drug levels of this patient showed concentrations in the toxic range. Consequently the dose was reduced.

4 Methods II: Dimensional Analysis

After selecting the time lag τ, we can construct the equivalence of the state space and search for the attractor of the system.

A chaotic motion in state space generally has a *fractal dimension*, a feature which allows its recognition. Periodic or quasiperiodic motions possess integer values (Fig. 7). The procedure widely used is the one proposed by Grassberger and Procaccia (1983), and Grassberger (1983): Within the reconstructed state space of dimension d_e, a reference point on the trajectory is chosen at random. Then the number of points which lie in a hypercube with radius r around this chosen point is counted. This counting is performed for subsequently larger radii r, until ultimately all points of the time series lie within this hypercube. In order to obtain a representative sample, the same procedure is repeated for other reference points. The counts are averaged for each radius r, and plotted against r using a double logarithmic scale. The resulting function has a $\int$-like shape. A linear fit is performed on the straight segment, the slope of which determines the desired "correlation dimension" d_c, provided the embedding dimension d_e is high enough. In order to test this supposition, the whole procedure must be repeated for successively higher embedding dimensions d_e until an independence of the calculated dimension d_c from the embedding dimension d_e is reached.

Let $N_{\vec{x}(i)}(r)$ be the number of adjacent points inside a hypercube of radius r which is centered around a reference point $\vec{x}(i)$ in the reconstructed phase space of a certain dimension d_e. The correlation dimension d_c will then be

$$d_c = \lim_{n_{data} \to \infty} \lim_{r \to 0} \frac{\log\left[\frac{1}{n_{ref}} \frac{1}{n_{data}} \sum_{i=1}^{n_{ref}} N_{\vec{x}(i)}(r)\right]}{\log r}$$

where n_{data} is the number of points in the reconstructed vectorial time series and n_{ref} is the number of reference points.[4]

Farmer et al. (1983) proposed an alternative method, called (averaged) *"pointwise dimension"* (d_{pw}) or *"information dimension."* A calculation of the dimension is done separately for every reference point. Then these estimations of the dimensionality are averaged:

$$d_{pw} = \lim_{n_{data} \to \infty} \lim_{r \to 0} \frac{1}{n_{ref}} \sum_{i=1}^{n_{ref}} \frac{\log\left[\frac{1}{n_{data}} N_{\vec{x}(i)}(r)\right]}{\log r}$$

Two points will be stressed here which suggest a preference for the latter method: Not all the reference points are equally well qualified for a good estimation since a chaotic attractor is generally nonuniform. If the straight region in the ∫-like curve becomes relatively small, the confidence of the estimation becomes low. Using the method of Farmer et al. (1983), such points can be easily excluded from the averaging process, thus enhancing the quality of estimation. The standard deviation of the dimensionality, for instance, will become smaller, as demonstrated by Holzfuß and Mayer-Kress (1986), only if a selection of the best reference points is used for the computations. The second point is also related to the nonuniformity of the attractor. If a reference point lies in a region with a low density of points, a relatively large radius r is needed to encompass most other points. On the other hand, a relatively short radius r will be sufficient if the reference point lies in a region with a high density of points. The effect is a vertical displacement of the ∫-like curves. In this case the averaging in the formula of d_c will result in a false, too flat slope, as demonstrated by Holzfuß and Mayer-Kress (1986). Furthermore, the averaged curve has a smaller scaling region and consequently the uncertainty in determining the slope increases. However, estimating each slope separately and then averaging across these estimates gives adequate accuracy. Only the averaged

---

[4] In order to be comparable with the subsequent dimension formula this formula may be written in a different form. Usually the original equation of Grassberger and Procaccia is cited as:

$$C(r) = \lim_{n_{data} \to \infty} \left\{ \frac{1}{n_{data}^2} \sum_{i,j=1}^{n_{data}} [\vartheta(r - |\vec{x}(i) - \vec{x}(j)|)] \right\}$$

with $\vartheta(x)$ as the Heavyside function [$\vartheta(x) = 0$, if $x \leq 0$, and $\vartheta(x) = 1$, if $x > 0$]. $|..|$ is the Euclidian norm or – as a simplification – the maximum norm.

Due to the so-called scaling law $C(r) \approx r^{d_c}$ which holds for small r, we end up with

$$d_c = \lim_{r \to 0} \frac{\log C(r)}{\log r}.$$

pointwise dimension takes into account the nonuniformity of the attractor with its various density of points in phase space.

Both methods give the same results when applied to analytic time series (Holzfuß and Mayer-Kress 1986). When applied to the EEG, however, dimensional values will become markedly different, as shown by Mayer-Kress and Holzfuß (1987).

We have used the method of averaged pointwise dimension for the present analyses, including only the best 20% of the reference points. First, 200 reference points were chosen at random, but only the best 40 points were kept for the determination of the pointwise dimension d_{pw}. By the best 40 points we mean those giving the relatively largest scaling regions, i.e., regions to which straight lines can be fit. This selection is made for every one of the embedding dimensions.

The embedding dimension d_e runs from 1 to 30. The dimensionality of the EEG, i.e., the dimension of an attractor, is determined by the asymptotic value reached by d_{pw} for larger embedding dimensions. (No saturation is obtained for purely random time series, i.e., noise, see Fig. 7).

5   Results for the Dimensional Analyses

An example of a selected EEG segment from a healthy, awake subject, with eyes open and closed, is illustrated in Fig. 4. It shows the pointwise dimension d_{pw} as a function of the embedding dimension d_e. The higher proportion of α-waves in the condition with eyes closed causes a decay in the dimensionality from about 12 to about 10. Figure 5 illustrates the results for a single patient. Whereas there seems to be no saturation in the eyes-closed condition, an attractor with a (fractal) dimension around 9 seems to dominate the EEG recorded while the subject held his eyes open. In contrast to the healthy control subject (Fig. 4), the patient (Fig. 5) failed to demonstrate clear α-waves in the eyes closed condition. Rather, the EEG became even more desynchronized.

We could not detect any systematic differences between patients and controls in the dimensionality of seizure-free EEG records. Generally, the dimensionality lies around 10. An exact calculation of such high dimensionalities requires much longer time series; therefore, these values should be interpreted only in relation to each other rather than in an absolute sense.

The EEG was also examined from a patient undergoing an absence seizure. The EEG segment, depicted in Fig. 6, shows slow waves, large amplitudes, and occasional spike patterns. The dimensional analysis yields lower values for the ictal than for the interictal period (Fig. 6).

In order to validate our computational strategies, we applied the analysis procedure to a random time series with the same length as the EEG segments. As shown in Fig. 7, there is no indication of saturation. d_{pw} climbs above 16 as d_e approaches 30. Comparisons of Figs. 4–6 with Fig. 7 demonstrate that there is a clear difference between a random, undetermined time series and the EEG, with its finite and probably fractal dimensions.

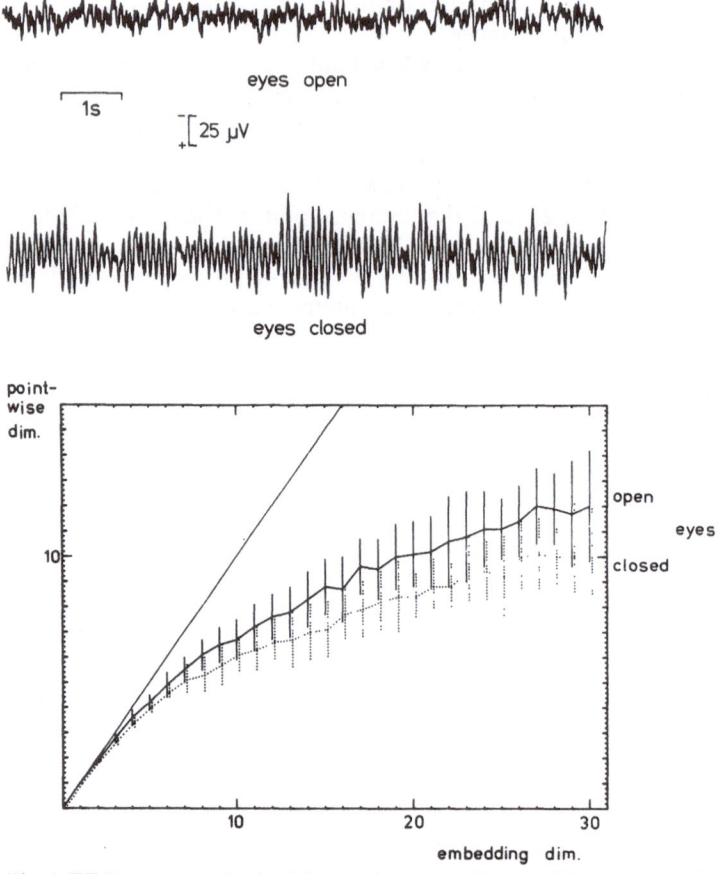

Fig. 4. EEG segments obtained from a healthy subject, as illustrated on the *top panel*, were analyzed to give the dimensional values illustrated in the *bottom panel*

5.1   Discussion

Dimensional analyses of the EEG have been presented before. Mayer-Kress and
Layne (1986) reported a fractal dimension of 8.7 for the waking EEG, which even
decreased to 5.1 after the subject kept his eyes closed and relaxed for a while. The
difference between eyes open and eyes closed conditions is very much in line with
our results for the healthy subjects. Our overall values, however, are higher than
those previously reported. The EEG segments chosen for the present analyses are
relatively long, which means that the time series could be less stationary. Or, the
EEG generating system might switch between different modes across extended
periods, exhibiting more of its complexity. It is also possible that the brain's regu-
lating systems which generate the EEG might work in simpler modes at times, but
switch to more complex modes when necessary. The relaxed waking state generat-

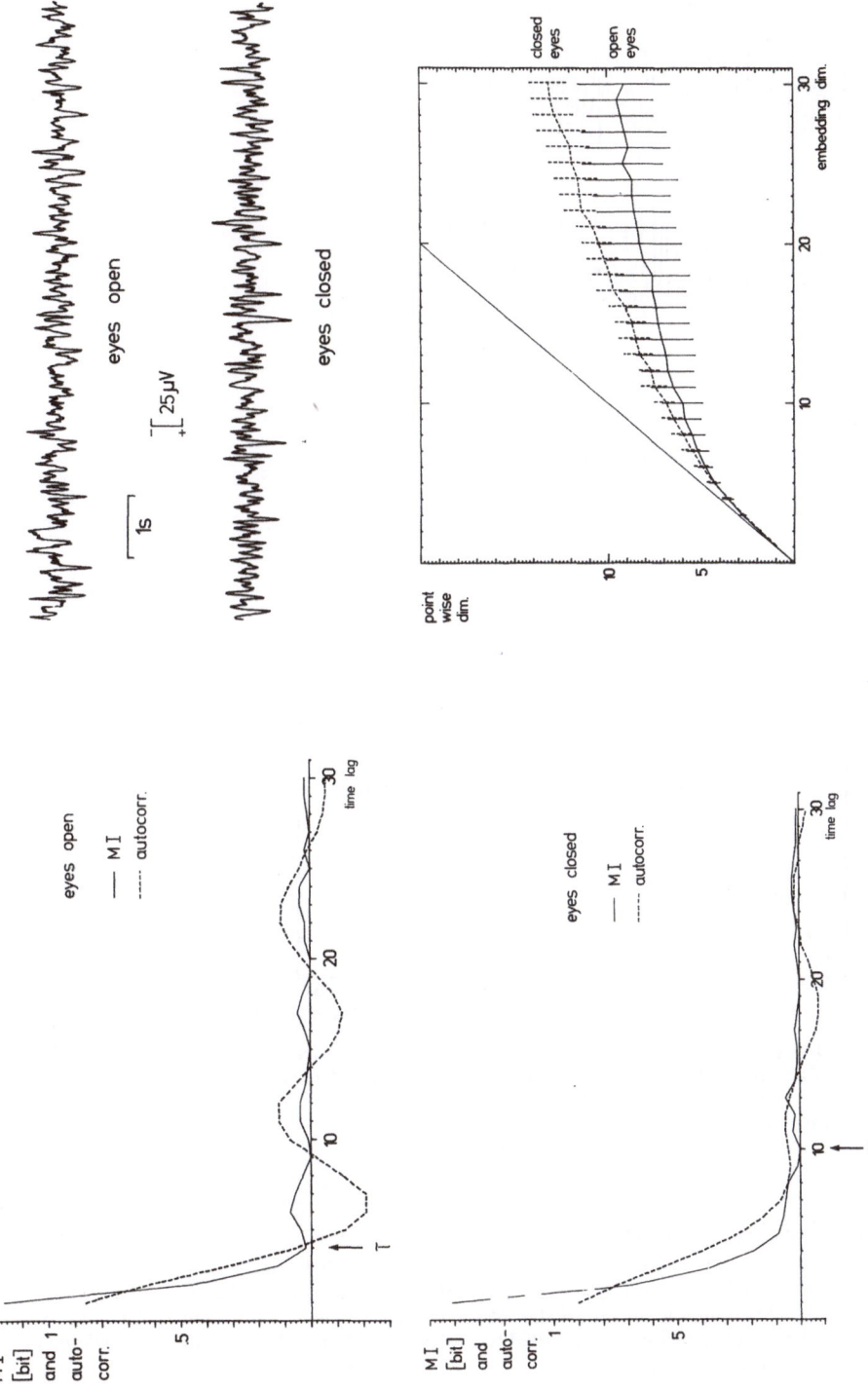

Fig. 5. Same analyses as in Fig. 4 for the EEG of an epileptic patient during a seizure-free period (*right side*). The panels on the *left side* show the mutual information function (*solid lines*) and the autocorrelation (*dotted lines*) for the illustrated EEG segments

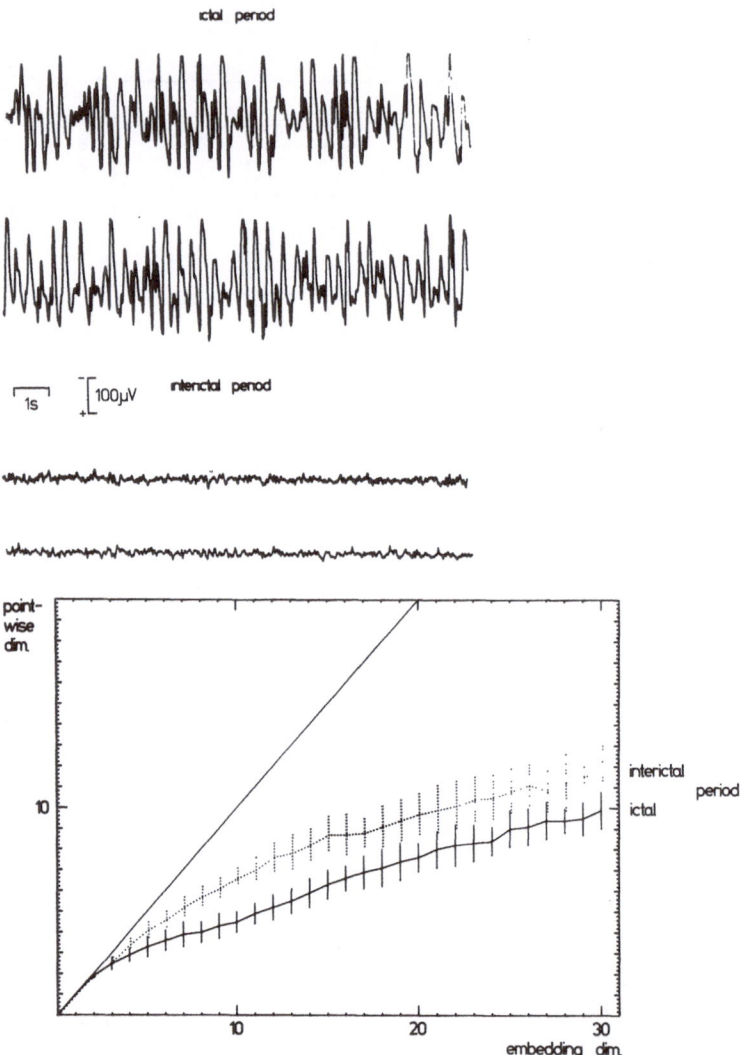

Fig. 6. Dimensional analyses of the EEG recorded during a seizure-free and during an ictal period of an epileptic patient. (Time series 4096 points.) The *upper panel* shows the segments of the raw EEG; The *lower panel* demonstrates the lower dimensionality of the EEG during the ictal period

Fig. 7. Same analysis as in Fig. 4 for noise, and for a five-dimensional quasiperiodic motion (5 d-torus). The time series for the torus is generated by $10 \sin(\sqrt{2}x) + 5 \sin(\sqrt{3}x) + 3 \sin(\sqrt{5}x) + 4 \sin(\sqrt{7}x) + 6 \sin(\sqrt{11}x)$. The overlapping periods are chosen not to be integral multiples of each other. Therefore, the dimensionality of the motion equals five, the number of component periods. (The results, the same as reported by Holzfuß and Mayer-Kress, confirm the adequacy of the procedures.)

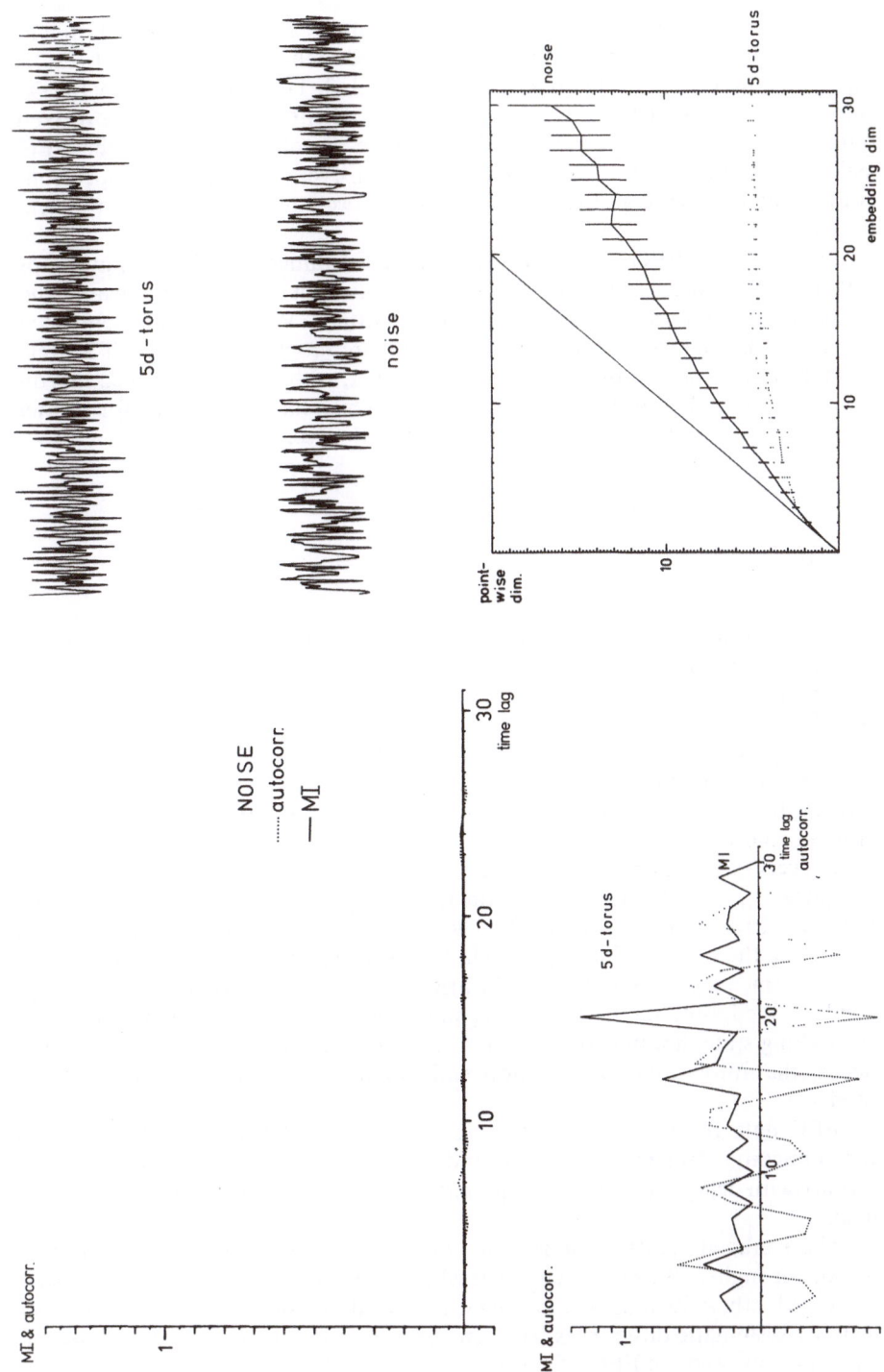

ing pronounced α-waves, seems to be such a "simpler mode" in terms of dimensionality. Of course, a condition with eyes closed could give a variety of results depending on what is going on in the subject's mind. This is illustrated by the example in Fig. 5: The EEG itself shows that this patient did not achieve a relaxed state with predominant α-waves. But, whereas visual inspection of the EEG makes it difficult to detect any difference between eyes closed and eyes open conditions, the dimensional analysis reveals the more complex mode for the eyes closed condition.

Epileptogenic EEG seems to display reduced dimensionality. An attractor with a fractal dimension between only 2 and 3 has been found by Babloyantz and Destexhe (1986) for an instance of a petit mal attack. We have found somewhat lower dimensional values during ictal (as compared to interictal) periods as well. A saturation of the pointwise dimension, however, with increasing embedding dimension, is not obvious (Fig. 6). One problem which has to be solved for further analyses is to obtain stationary, extended EEG time series.

6   Summary

Research on nonlinear systems theory has proven to be essential for the understanding of biological systems. The present contribution examines the usefulness of nonlinear system theory for the description of electrical brain activity. The issue addressed is how to characterize the system's (fractal) dimension. Grassberger and Procaccia (1983) have suggested a "correlation dimension", and Farmer et al. (1983) a "pointwise dimension." The latter seems to be preferable and therefore has been applied in the present paper (comparison of these two quantities in Holzfuß and Mayer-Kress 1986).

In every case, an optimal time lag must first be chosen in order to reconstruct the phase space. The autocorrelation function and the "mutual information" function (Fraser and Swinney 1986) were compared for their effectiveness in evaluating the optimal time lag. The latter has been used extensively in previous research since it accounts for nonlinearities. Nevertheless, for the 22 EEG-segments studied, both methods yielded equal results. We conclude that, at least for the waking state, nonlinearities do not play a major role in EEG sequences, and hence, the autocorrelation function with its shorter calculation time may be used.

EEG data from seven epileptic patients and two healthy controls indicated that anticonvulsant medication expands the optimal time lag. This is in conformity with delayed event-related potentials and reaction times in medicated patients.

The averaged pointwise dimension of the waking EEG was typically found to be nine or higher. If α-waves emerged while the eyes were closed, the dimension decreased, otherwise it increased. Regarding the dimensionality, we could not detect any systematic differences between the seizure-free EEG recorded from epileptic patients and the EEG of controls. As expected, the EEG during an absence seizure is characterized by lower values than the one recorded during interictal periods.

Acknowledgments. We appreciate the helpful comments of Drs Brigitte Rockstroh, Walton R. Roth, and Otto E. Rössler. This work was supported by grants from the Max Kade Foundation and the Deutsche Forschungsgemeinschaft (SFB 307).

References

Albano AM, Abraham NB, de Guzman GC, Tarroja MFH, Bandy DK, Gioggia RS, Rapp PE, Zimmerman ID, Greenbaun NN, Bashore TR (1986) Lasers and brains: complex systems with low-dimensional attractors. In: Mayer-Kress G (ed) Dimension and entropies in chaotic systems. Springer, Berlin Heidelberg New York, pp 231–252

Babloyantz A (1985) Strange attractors in the dynamics of brain activity. In: Haken H (ed) Complex systems – operational approaches in neurobiology, physics, and computers. Springer, Berlin Heidelberg New York, pp 116–122

Babloyantz A, Destexhe A (1986) Low-dimensional chaos in an instance of epilepsy. Proc Natl Acad Sci USA

Babloyantz A, Salazar JM, Nicolis C (1985) Evidence of chaotic dynamics of brain activity during the sleep cycle. Phys Lett 111(3):152–156

Bülow I v., Elbert T, Lutzenberger W, Rockstroh B, Birbaumer N, Canavan A (1988) Effects of hyperventilation on EEG-frequency and potentials in epilepsy. J Psychophysiol (in press)

Crutchfield JP, Farmer JD, Packard NH, Shaw RS (1986) Chaos. Sci Am 255(6):46–57

Dvorak J, Siska J (1986) On some problems encountered in the estimation of the correlational dimension of the EEG. Phys Lett 118 A(2):63–66

Elbert T (1987) Regulation corticaler Erregbarkeit – Im EEG ein deterministisches Chaos? In: Weinmann HM (ed) Zugang zum Verständnis höherer Hirnfunktionen durch das EEG. Zuckschwerdt, München, pp 93–107

Elbert T, Rockstroh B (1987) Threshold regulation – a key to the understanding of the combined dynamics of EEG and event-related potentials. J Psychophysiol 1:317–333

Farmer JD, Ott E, Yorke JA (1983) Dimension of chaotic attractors. Physica 7D:153–180

Fraser AM (1986) Using mutual information to estimate metric entropy. In: Mayer-Kress G (ed) Dimension and entropies in chaotic systems. Springer, Berlin Heidelberg New York, pp 82–91

Fraser AM, Swinney HL (1986) Independent coordinates for strange attractors from mutual information. Phys Rev A 33(2):1134–1140

Glass L, Shrier A, Bélair J (1986) Chaotic cardiac rhythms. In: Holden AV (ed) Chaos. – Nonlinear science: theory and application. Manchester University Press, Manchester, pp 237–256

Graf KE (1986) Grundlagen der experimentellen Anwendung nichtlinearer dynamischer Systemtheorie. Universität Tübingen

Grassberger P (1983) Generalized dimensions of strange attractors. Phys Lett 97 A(6):227–230

Grassberger P, Procaccia I (1983) Characterization of strange attractors. Phys Rev Lett 50(5):346–349

Holzfuß J, Mayer-Kress G (1986) An approach to error estimation in the application of dimension algorithms. In: Mayer-Kress G (ed) Dimension and entropies in chaotic systems. Springer, Berlin Heidelberg New York, pp 114–122

Lorenz EN (1963) Deterministic nonperiodic flow. J Atmos Sci 20:130–141

Mayer-Kress G, Holzfuß J (1987) Analysis of the human electroencephalogram with methods from nonlinear dynamics. In: Rensing L, an der Heiden U, Mackey MC (eds) Temporal disorder in human oscillatory systems. Springer, Berlin Heidelberg New York

Mayer-Kress G, Layne SP (1986) Dimensionality of the human electroencephalogram. Submitted to Proceedings of New York Academy of Sciences conference on perspectives in biological dynamics and theoretical medicine, Bethesda

Packard NH, Crutchfield JP, Farmer JD, Shaw RS (1980) Geometry from a time series. Phys Rev Lett 45(9):712–716

Rapp PE (1986) Oscillations and chaos in cellular metabolism and physiological systems. In: Holden AV (ed) Chaos – nonlinear science: theory and application. Manchester University Press, Manchester, pp 179–208

Rapp PE, Zimmerman ID, Albano AM, de Guzman GC, Greenbaun NN (1985) Dynamics of spontaneous neuronal activity in the simian motor cortex: the dimension of chaotic neurons. Phys Lett 110 A(6):335

Rapp PE, Zimmerman ID, Albano AM, de Guzman GC, Greenbaun NN, Bashore TS (1986) Experimental studies of chaotic neuronal behaviour: cellular activity and electroencephalographic signals. In: Othmer HG (ed) Nonlinear oscillation in biology and chemistry. Springer, Berlin Heidelberg New York, pp 175–205

Rockstroh B, Elbert T (1988) On the regulation of excitability in the cerebral cortex – a bridge between EEG and attention? In: Pickenhain L (ed) Cortical DC-potential shifts and human performance (in press)

Rockstroh B, Elbert T, Lutzenberger W, Altenmüller E, Diener HC, Birbaumer N, Dichgans J (1987) Effects of the anticonvulsant carbamazepine on event-related brain potentials in humans. In: Barber C, Nodar H (eds) Evoked potentials III. Butterworth, London, pp 361–369

Rössler OE (1983) The chaotic hierarchy. Z Naturforsch 38 a:788–801

Roux JC, Simoyi RM, Swinney HL (1983) Observation of a strange attractor, Physica D 8:257–266

Schaffer WM, Kot M (1986) Differential systems in ecology and epidemioloy. In: Holden AV (ed) Chaos –nonlinear science: theory and application. Manchester University Press, Manchester, pp 158–178

Swinney HL, Gollub JP (1986) Characterization of hydrodynamic strange attractors. Physica 18 D:448–454

Takens F (1981) Detecting strange attractors in turbulence. In: Rand DA, Young LS (eds) Lecture notes in mathematics 898. Springer, Berlin Heidelberg New York, pp 366–381

Whitney H (1936) Differentiable manifolds. Ann Math 37:645

Analysis of Strange Attractors in EEGs with Kinesthetic Experience and 4-D Computer Graphics*

W. J. FREEMAN

1 Introduction

Newly developed techniques for numerical analysis of the time series derived from physical and chemical systems displaying turbulence and other seemingly random behavior have recently been applied to electroencephalographic potentials (EEGs). Estimates for the lower bounds of measures of the fractal dimension, Hausdorff dimension, Lyapunov exponents, Kolmogorov entropy, and other related coefficients have indicated that by suitable means of observation the background "noise" generated by brains may be nonrandom, and may be constrained by as yet undefined aspects of neural mechanisms generating it. In a word, brain activity, at least in some states and in some brain parts, appears to be chaotic and not stochastic (Rapp et al. 1985; Freeman and Viana Di Prisco 1986; Babloyantz and Destexhe 1986; Havstadt and Ehlers 1987; Meyer-Kress 1987; Freeman 1987a, b, 1988).

This inference may have profound consequences for theories of brain function, leading to better understanding both of the neural mechanisms generating chaos, and the purposes that it may serve in the genesis of goal-directed behavior (Skarda and Freeman 1987). In order to exploit this new opportunity we need to know much more about the dynamics of neural masses that underlies the appearance of the EEGs recorded in behaving animals and man.

2 Methods

One way to study the dynamics is to plot the set of points in a time series (an electroencephalographic or EEG trace as exemplified in Fig. 1) in a coordinate space, in which one axis is the amplitude at each time t, and each other axis (Fig. 2) is the amplitude at a time lag $t+nT$, where n is an integer 1, ... up to a desired number of axes (a Ruelle plot) and T is a time increment usually expressed as a multiple of a digitizing interval. The figure will appear as a cloud of points that may be rotated for projection or section into any plane in the space for display. It is convenient to connect successive points with a line segment for display as a continuous trajectory. When it is plotted in 3-D or a higher dimension the figure may be rotated and translated to inspect its shape.

A variant on this approach is to record EEGs at two or more sites and use these rather than one trace repeatedly lagged. Yet another is to rotate the coor-

* Originally published in Başar E, Bullock TH (eds) Brain dynamics. Springer, Berlin Heidelberg New York, pp 512–520 (Springer series in brain dynamics, vol 2). Cross references refer to that volume.

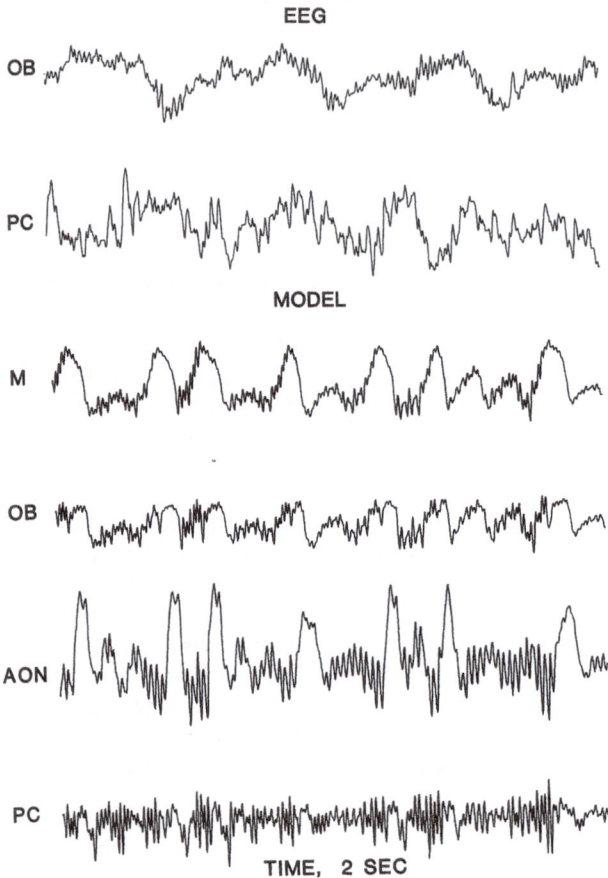

EEG

OB

PC

MODEL

M

OB

AON

PC

TIME, 2 SEC

Fig. 1. Examples of EEG
activity recorded from a
waking rat from the olfactory bulb (*OB*) and prepyriform cortex (*PC*), and of
simulated EEGs from a
model of the olfactory system including outputs from
the mitral cell population
(*M*) and the anterior olfactory nucleus (*AON*). (From
Freeman 1987 b)

dinate axes by singular value decomposition or factor analysis, so that the axes
are orthogonal for plotting the *n*-dimensional distribution of the variance (Mees
et al. 1987).

A successful outcome of these procedures is the detection of spatial structure
in the projections into one or more planes of section. The cloud of points or line
segments manifests "flow" of "activity" in the system, that may expand in some
directions and contract in others, and that may be folded back into itself to form
loops and whorls. These structural features are manifestations to be expected of
the constraints or invariants of the dynamics that generates the neural activity.

If the duration of the time series, the temporal digitizing interval, and the time
lags and recording locations are all varied and ultimately optimized to enable the
extraction of these visual features, then rotation and translation of the cloud or
of the position of observation should convey to an observer a perceptual experience that is reproducible over repeated sample traces of the EEGs from the same
or similar subjects in the same or similar states. The "object" is then communi-
cable by a photograph, sketch, or cinema.

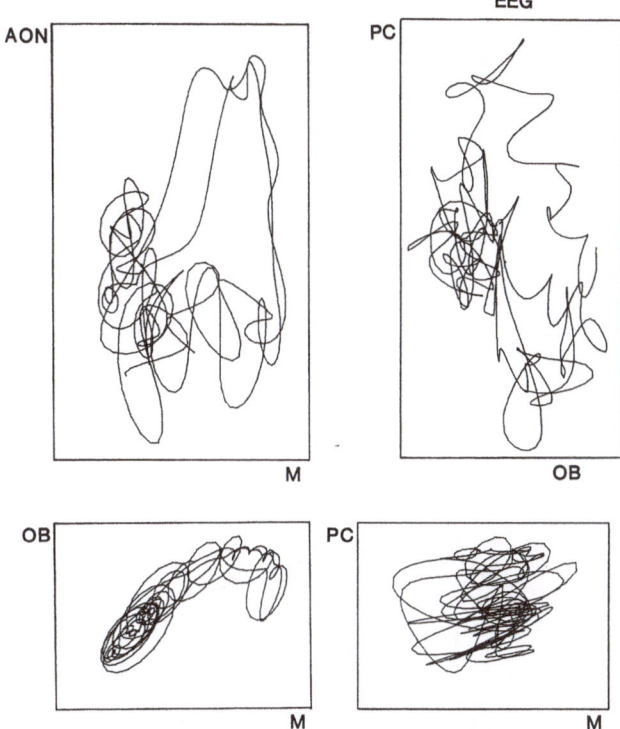

Fig. 2. Examples of tangles generated by plotting one variable against another for the EEG (*upper right frame*) and the outputs of the model. The dimension estimated by the Grassberger and Procaccia (1983) and Guckenheimer (1984) methods averaged 5.46 for the model and 5.92 for the EEG (Freeman 1988). (From Freeman 1987 b)

3   Modes and Models

These prescriptions are of little use with the conventionally recorded scalp EEGs, which tend to consist of potential fields from unknown numbers of neural "generators" all summing in the volume conductor of the brain and scalp tissues, and over time epochs that in the waking or conscious condition will include indefinite numbers of state transitions subserving the ongoing evolution of goal-directed behavior. Contributions from electrode noise, muscle potentials and movement artifacts may serve further to inflate the overall variability of the data. The variability may be arbitrarily reduced by improper use of smoothing and detrending filters, and by use of an excessively small digitizing interval or an overly brief duration of recording epoch (Albano et al. 1986). Examples of Ruelle plots of EEG data typically lack interesting and reproducible structure and resemble bowls of spaghetti.

What is needed is some prescription on what look for by way of structure. In our experience this requirement is best met by constructing models of the dynamics of neural masses, which consist of sets of coupled nonlinear ordinary differential equations (ODEs), for which the solutions generate time series and spatial patterns that simulate the corresponding features of EEGs. The most useful models are those that closely hew to the known anatomical connections and open-

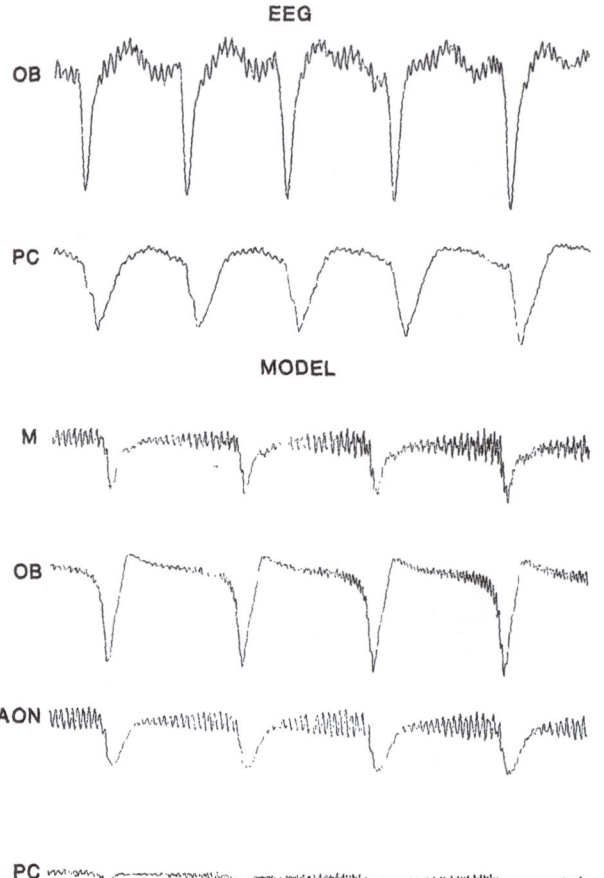

Fig. 3. Examples of seizure activity from the rat EEG and from the model. The correlation dimension of the output of the model averaged 3.76 and that of the EEG was 2.52 (Freeman 1988). (From Freeman 1987b)

loop properties of component neurons (Freeman 1975, 1987a, b; Freeman and Skarda 1985; Baird 1986).

Our best example is a model of the olfactory system that suffices to generate chaotic activity that with appropriate parameter settings is not only indistinguishable from low-pass filtered random activity (so-called colored noise as distinct from white noise) by all conventional statistical tests but is also indistinguishable from olfactory EEGs at various sites, whether seen in time series or Ruelle plots (Figs. 1 and 2). The model consists of 30 first-order ODEs, but the dynamic elements representing local subsets of neurons are formed into four main groups with delays between them. This leads to the hypothesis that the dynamics modes, degrees of freedom, or in some sense "dimension" of this system is on the order of 4, not 30, and that the number of modes in the corresponding olfactory system is less than the numbers of nuclear clusters of cells or glomeruli (which are not readily defined) or the numbers of neurons (which may well exceed 100 million).

This hypothesis was evaluated by applying several numerical algorithms to the real and simulated EEGs from the "normal" basal state. The results have shown that the two sets of estimates of correlation dimensions are equal to each other within the standard error of measurement, and have suggested that they lie between 5 and 6 (Freeman 1988).

This part of the brain has the capability for generating an abnormal EEG pattern of partial seizure that is accompanied in animals by a form of epilepsy (Fig. 3). Experimental analysis shows that its onset is occasioned by a catastrophic "structural" failure, in that a synaptic connection fails between two of the main parts of the system. This "disconnection" can be construed to reduce the number of modes to three, with the expectation that the fractal dimension is reduced to between two and four. The same abnormal pattern is generated by the model when it is similarly disconnected (Freeman 1986). Our estimates of fractal dimensions by the Grassberger and Procaccia (1983) and the Guckenheimer (1984) methods show that our prediction of four dimensions was too low (Grajski and Freeman 1988) but that there is agreement between values for the EEG and the model, both giving values of five to six dimensions.

4   3-D and 4-D Displays

This condition of the "disconnected" state yielding seizure activity gives a highly structured cloud of points and connected line segments that, from various perspectives in Euclidean space, can be seen to conform to a filled-in 2-torus (Freeman 1987 b). Analysis of the model shows that this pattern of activity can be reached by the Ruelle-Takens quasiperiodic route to chaos, starting from a point attractor and passing through two Hopf bifurcations before the phase transition to chaos. The structure can be construed to be a chaotic attractor, because in the model its size and shape is independent fo the path by which it is reached from an arbitrary starting point, although the form of the time series of the activity is strongly dependent on the initial conditions (Freeman 1986). The same conclusion holds for the animals because the seizure can be repeatedly induced with substantially the same time course to recovery. Within the seizure state the time course is not altered by transient perturbations induced by superimposed electrical stimulation, but no two time series are identical. This example is felicitous, because it has a chaotic attractor that can readily be embedded and examined in normal 3-D space (Fig. 4). It is a good test object for improving our methods of data processing and display. However, a major problem emerges when the animal recovers from the seizure. Typically there is an abrupt return to normal behavior and a concomitant shift of the EEG back to its normal pattern. The Ruelle plots of the EEGs in this normal condition reveal a structure that is both interesting and reproducible, but it cannot yet be identified with or encompassed within any known geometric form. The simulated EEG as well yields structure in its phase portraits that is beautiful to behold but baffling. Both portraits manifest chaotic attractors, because they are stable under perturbation and recur from differing starting conditions at the time of transition into these reproducible states.

Our analysis of the model and our measurements on the fractal dimensions of the real and artificial normal EEGs lead us to believe that the attractor and

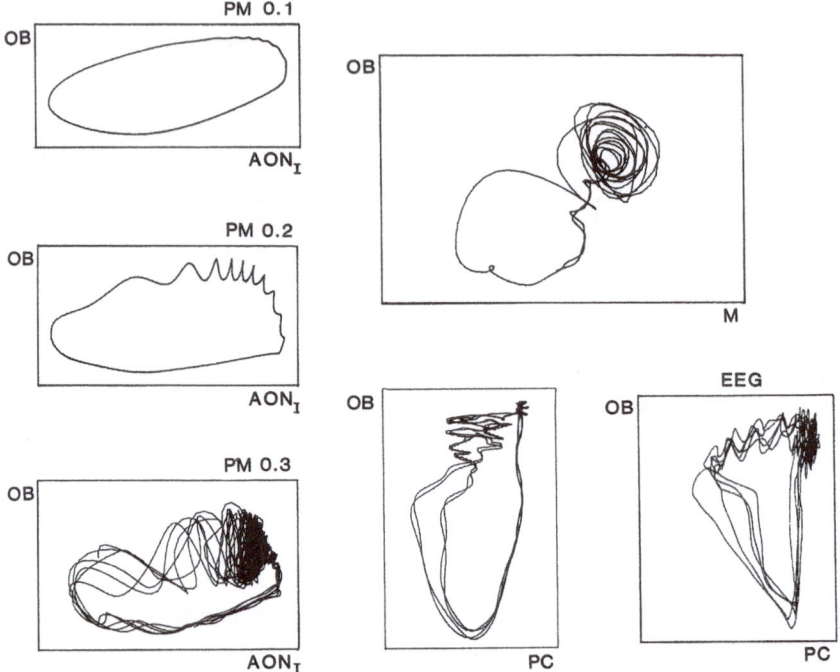

Fig. 4. Examples of phase plots of the EEG (*lower right*) and of the model. The series at *left* shows the sequence of two Hopf bifurcations in the model starting from a point attractor, culminating in the onset of chaos with increase in a bifurcation parameter *PM,* a synaptic connection strength in the model. (From Freeman 1987b)

its simulacrum exist in 4-D or yet higher space. It is incumbent on us, therefore, to postulate a 4-D structure and deduce what 3-D geometric forms will appear when planar sections are taken through it. This procedure has been followed in constructing images of the 4-D hypercube, which in 3-D sections give cubes, octahedrons, and other regular solids (Banschoff 1984, 1986), and the 4-D hypersphere, which yields spheres and 2-tori in various projections (Banschoff 1986). A promising candidate for the normal basal EEG attractor is the filled-in 3-torus (hypertorus), because the model contains three coupled oscillators with incommensurate characteristic frequencies (Freeman 1986). One of the geometric forms that is predicted for a planar section through a 3-torus is a 2-torus, and some of the "views" we have of the attractor in three dimensions suggest the appearance of a 2-torus. Kaneko (1986) has shown that this structure also provides a route into chaos. However, there is not yet available an intensive study of this structure on which to base our expectations on what to seek, or to tell us how to define and characterize the features that will support or exclude the 3-torus as a hypothesis.

5   Kinesthetics and the EEG

The major experimental and conceptual problem we face now is how to explore the geometry of 4-D structures with the aid of computer graphics. A solution proposed by Rössler (1978, 1981) is to "fly the attractor" with the aid of a flight simulator. An example is given by H.-O. Peitgen's film (1985), "Fly Lorenz," for this famous strange attractor that can be visualized in 3-D. Experience has shown that an observer can rapidly become familiar with a 4-D object, when he is given the ability to move his position with respect to the object in the same way that an aircraft pilot moves his point of observation with respect to his visual world.

This experience is in close conformity to our understanding of how our visual nervous system operates. Our retinal input is restricted to 2-D images. From these we construct 3-D conceptions of our world, and the essential means is to move ourselves through the world and to move objects in that world by use of our limbs (Poincare 1958, 1963). Locomotion by voluntary goal-directed muscular coordination is the key to 3-D visual perception. There is good reason to suppose that our nervous systems can adapt to 4-D in much the way that they do to 3-D. As O. E. Rössler (personal communication) points out, experiencing a hypercube is like entering a house with eight rooms. Even in our 3-D world we must experience the rooms one at a time and build our conception by walking through them serially. The operator at a graphics terminal, if the visual feedback from his movements is fast enough to be "realistic," can quickly become at home in a hypercube, but someone watching the video screen over his shoulder will not know what is going on. Numerous studies in visual psychophysics have anticipated this result and strongly support this interpretation.

6   Implications

This convergence of neurophysiology, mathematics, and computer science holds great promise for all three fields. For the physiologist the immediate application is in the exploration of the hypergeometry of neural strange attractors revealed by EEGs, with the hope of deducing how they both shape and express the underlying neural dynamics, the ways in which they are stable, how they change with behavioral state, how pathological attractors can develop, and what remedial actions can be taken to reestablish normal dynamics. The use of computer graphics in 4-D or possibly in higher spaces may be the most important technical innovation in electroencephalography since the invention of the ink-writing polygraph as a clinical diagnostic tool.

The insights derived from our experience with 4-D perception may provide new ways to reformulate our ideas about visual perception and concept formation (Rucker 1984). The scientists most likely to benefit may be those who are intent on building systems and devices that can "see" in the real world and control their behaviors in 3-D space. The physiologist and psychophysicist already understand that perception is the incorporation of external information into the nervous system by movements into the world. Cognitive scientists place too great emphasis on the response to stimuli as distinct from the grasping of stimuli. If one desires a machine to read, one must first make it to write and to read what it has written. If one wishes to fly a plane, one must fling one's body about the sky with one's hands on the controls, until the machine becomes a part of the self and one's body image extends to the wingtips and the tail; reading about it will not suffice.

Mathematicians have made substantial progress in the past decade in their understanding of systems with three equations, but still little has been done with sets of four, which are the threshold for the domain of "hyperchaos" (Rössler 1979, 1983). The barrier seems to be in the inuitive apprehension of ("grasping at") the hypergeometric structures from the phase portraits of these systems. The ready access to these experiences through kinesthetics and computer graphics may open a field of new human endeavor in hypergeometry as broad as that of the clinical applications for the EEG.

Finally, the technology of computer graphics is already rich with new art forms (Schwenk 1976; Peitgen and Richter 1986). As cinema put new dimension into photography, and sound recording added another to that, the fusion of visual, reafferent and proprioceptive images by use of a graphics terminal and a mouse in each hand (and perhaps under each foot as well – this arrangement would be familiar to an experienced glider pilot) may open wide a door that has been sought by scientists and philosophers for a century or more (Abbott 1952). The cost of the necessary equipment at present is far too high for general availability. Those machines that are fast enough and have the capacity for handling the large numbers of data points needed have their microcode written in firmware for 3-D. Various tricks and strategems to circumvent this design limitation have their drawbacks, but the realization of full 4-D or even 5-D capability is a matter of cost and demand. We can expect within a short time to witness an explosive development with many as yet unforeseen new applications in "applied hypergeometry." Certainly our bodies are limited in dimensions but our minds are not.

Acknowledgment. This work was supported by a grant MHO6686 from the National Institute of Mental Health, United States Public Health Service.

References

Abbott EA (1952) Flatland, 6th edn. Dover, New York
Albano AM, Mees AI, deGuzman GC, Rapp PE (1986) Data requirements for reliable estimation of correlation dimensions. In: Holden AV (ed) Chaotic biological systems. Pergamon, New York
Babloyantz A, Destexhe A (1986) An instance of low-dimensional chaos in epilepsy. Proc Natl Acad Sci USA 83:3513–3521
Baird B (1986) Nonlinear dynamics of pattern formation and pattern recognition in the rabbit olfactory bulb. Physica 22D:150–175
Banschoff TF (1984) The hypercube (16 mm color film) International Film Bureau, Chicago
Banschoff TF (1986) Visualizing two-dimensional phenomena in four-dimensional space: a computer graphics approach. In: Wegman EJ, DePriest DJ (eds) Statistical image processing and graphics. Dekker, New York, pp 187–202, 18 plates
Freeman WJ (1975) Mass action in the nervous system. Academic, New York

Freeman WJ (1986) Petit mal seizure spikes in olfactory bulb and cortex caused by runaway inhibition after exhaustion of excitation. Brain Res Rev 11:259–284

Freeman WJ (1987a) Techniques used in the search for the physiological basis of the EEG. In: Gevins A, Remond A (eds) Handbook of EEG and clinical neurophysiology, vol 3 part 2. Elsevier, Amsterdam, chap 18

Freeman WJ (1987b) Simulation of chaotic EEG patterns with a dynamic model of the olfactory system. Biol Cybern 56:139–150

Freeman WJ (1988) Strange attractors that govern mammalian brain dynamics shown by trajectories of electroencephalographie (EEG) potential. IEEE Trans Circuits and Systems 35:781–784

Freeman WJ, Skarda CA (1985) Spatial EEG patterns, nonlinear dynamics and perception: the neo-Sherringtonian view. Brain Res Rev 10:147–175

Freeman WJ, Viana Di Prisco G (1986) EEG spatial pattern differences with discriminated odors manifest chaotic and limit cycle attractors in olfactory bulb of rabbits. In: Palm G, Aertsen A (eds) Brain theory. Springer, Berlin Heidelberg New York, pp 97–119

Gleick J (1987) Chaos. Making a new science. Viking, New York

Grassberger P, Procaccia I (1983) Measuring the strangeness of strange attractors. Physica 9D:189–208

Guckenheimer J (1984) Dimension estimates for attractors. Contemp Math 28:357–367

Havstadt JW, Ehlers CL (1987) Attractor dimension for small data sets. Preprint, La Jolla CA

Kaneko K (1986) Collapse of tori and genesis of chaos in dissipative systems. World Scientific, Singapore

Mees AI, Rapp PE, Jennings LS (1987) Singular value decomposition and embedding dimension. Physical Rev A36:340–346

Meyer-Kress G (1987) Application of dimension algorithms to experimental chaos. In: Hao Bai-Lin (ed) Directions in chaos, World Scientific, Singapore, pp 122–147

Peitgen H-O (1985) Fly Lorenz (16 mm color film). International Film Bureau, Chicago

Peitgen H-O, Richter P (1986) The beauty of fractals. Images of complex dynamical systems. Springer, Berlin Heidelberg New York

Poincare H (1958) The value of science. Dover, New York, pp 37–74

Poincare H (1963) Last essays. Dover, New York, pp 25–44

Rapp PE, Zimmerman ID, Albano AM, deGuzman GC, Greenbaum NN, Bashore TR (1985) Experimental studies of chaotic neural behavior: cellular activity and electroencephalographic signals. In: Othmer HG (ed) Nonlinear oscillations in biology and chemistry. Springer, Berlin Heidelberg New York, pp 175–205

Rössler OE (1978) Deductive biology – some cautious steps. Bull Math Biol 40:45–58

Rössler OE (1979) Chaos. In: Guttinger W, Eikemeier H (eds) Structural stability in physics. Springer, Berlin Heidelberg New York, pp 290–309

Rössler OE (1981) An artificial cognitive-plus-motivational system. Prog Theor Biol 6:147–160

Rössler OE (1983) The chaotic hierarchy. Z Naturforsch 38a:788–801

Rucker R (1984) The fourth dimension. Houghton-Mifflin, Boston

Schwenk T (1976) Sensitive chaos. Schocken Books, New York; Steiner, Dornach

Skarda CA, Freeman WJ (1987) How brains make chaos in order to make sense of the world. Brain Behav Sci 10:161–195

Chaos in Brain Function and the Problem of Nonstationarity: A Commentary*

G. J. Mpitsos

1 Introduction and Historical Perspective

Over 16 years ago, the work of two of the participants in the present conference clearly pointed to the need to consider the brain as a noisy processor (Adey 1972) in which statistical mechanisms (John 1972) lead to the production of organized behavior. The difficulty in pursuing this problem further in terms of the individual neurons that comprise the network has been in the lack of conceptual tools with which to understand the informational language arising from simple neuroanatomic structures such as the convergence of two neurons onto a third (Mpitsos et al. 1978), or, as Sperry (1981) has stated it, in our inability to handle the "three-bodies problem." Bullock (1984) has observed that "circuit analysis," the once great hope of neurobiology, will not by itself "provide the major insights necessary to understand the emergent mechanisms present in complex systems," even if these systems are as simple as the stomatogastric ganglion of the lobster (Selverston 1980). The realization of this fact itself is perhaps a major achievement in invertebrate neurobiology. Nonetheless, while major strides have been made in understanding the self-organizing processes in distributed, parallel networks (e.g., Amit et al. 1985; Grossberg 1980; Hopfield 1982; Jeffrey and Rosner 1986; Kleinfeld and Sompolinsky 1987; Pellionisz and Llinas 1983; Rumhelhart et al. 1986; Sejnowski and Rosenberg 1987; Werbos 1974), the subject of noise has not been broadly addressed, and only recently has it received rekindled interest.

This interest has come about through the analysis of the dynamics in spontaneous and stimulus-evoked neural responses, and through observations of behaviors and of the neural patterns that generate them. From the application of dynamical techniques, it seems now that chaos may be involved in sleep and seizure EEGs (Babloyantz and Destexhe 1986; Babloyantz et al. 1985), baseline olfactory bulb EEGs (Freeman and Skarda 1985; Skarda and Freeman 1987), bulbar EEGs in response to odors (Skinner this volume) the responses of electrically driven neurons (Aihara and Matsumoto 1986; Hayashi and Ishizuka 1986), spontaneous firing of cortical neurons (Rapp et al. 1985), and in computer simulation of single neuron and network properties (Chay 1985; Labos 1986).

From studies of behavior, interest in variability has arisen in human studies which show that certain types of fluctuations in hand and finger movements may provide a mechanism for phase transitions to occur (Kelso et al. 1986). In our own work, observations of extensive variability in behaviors, motor patterns, re-

* Originally published in Başar E, Bullock TH (eds) Brain dynamics. Springer, Berlin Heidelberg New York, pp 521–535 (Springer series in brain dynamics, vol 2). Cross references refer to that volume.

sponses of single neurons, and in their interrelated responses has led us to propose that variability may not be simply a product of supposed inexactness of biological systems, but, rather, that it may be an essential feature by which neural activity self-organizes (Cohen and Mpitsos 1983 a, b; Mpitsos and Cohan 1986 a, b). State space analyses of the dynamics in such motor patterns have supported this assertion by showing that the responses of key neurons in the generation of motor patterns may be governed by chaotic attractors (Mpitsos et al. 1988 a, c). Moreover, as has been proposed for the transition between chaos and limit cycles (Freeman and Skarda 1985; Freeman and Viana Di Prisco 1986; Skarda and Freeman 1987), we have suggested that chaos and other forms of variability may provide a means by which activity can shift from one motor pattern to another. Sources of variations may also be useful in optimization of network activity (Burton and Mpitsos 1988 a, b; Kirkpatrick et al. 1983; Jeffrey and Rosner 1986).

The above findings permit us to answer in the affirmative Adey's (1972) question: "Is it possible to envisage an information-processing system in which the very presence of on an ongoing noise-like activity produced no degradation in information-handling ability, and might even enhance it?" We can go further and build on Adey's insight to say that while there may be many forms of noise in brain function, among which we include variations having white spectra, or even invariant perturbations, variations arising from chaos cannot be separated from the informational content of the signal because they themselves are an integral part of the information and are generated by mechanisms that need not contain explicit elements of noise.

In the remainder of this chapter, I shall address the subjects of chaos, difficulties in assessing it in behavioral systems, the problem of nonstationarity, and the ability of neural networks to read and transmit chaos. Inasmuch as the intent of this chapter is to provide a commentary on the implications of chaos, and on difficulties in obtaining evidence for it, rather than an exposition of the findings themselves, I refer readers to the publications listed above for details on the specifics of the experimental findings and for the application of dynamical theory to their analysis. Further introductions and applications of dynamical theory may be found in Abraham and Shaw (1983); Babloyantz (this volume); Freeman (this volume; Garfinkel and Freeman (this volume); Gleick (1987); Shaw (1986); and in the volumes edited by Holden (1986) and Degn et al. (1986).

2   Duration Multifunctionality, and Nonstationarity of Coordinated Activity

2.1   Experimental Preparation

The experimental foundation, which we use as a point of departure to address broader issues in the following discussions, stems from observations of whole animal behavior, from motor patterns recorded in relatively intact preparations (e.g., McClellan 1982 a; Mpitsos and Cohan 1986 a; Mpitsos et al. 1978), and from the responses of individual neurons that take part in generating integrated output of activity in isolated (deafferented) nervous systems (Cohan and Mpitsos 1983 a, b; McClellan 1982 b; Mpitsos and Cohan 1986 b; Mpitsos et al. 1988 c).

Our experimental animal is the carnivorous marine mollusc *Pleurobranchaea californica,* a snail-like animal having no shell. As in humans, the most complicated behaviors of this animal involve its mouth, lips, jaws, and tongue. These behaviors include several components of feeding and regurgitation, rejection, which resembles a reverse of the sequential bite-ingestion movements of feeding, defensive biting, and self- and interanimal gill-grooming. (For photographs of feeding behavior see Mpitsos et al. 1978.) Although these behaviors produce obviously different effects (consider the difference between ingestion and regurgitation), their repetitive movements are quite similar to one another.

The nervous system of *Pleurobranchaea* contains about 10000 neurons, several thousand of which may become active in a given behavior, but the crucial information for all coordinated buccal-oral behaviors must pass through and is generated by only about 30 neurons. We have examined the activity of individual neurons of this group and of the motoneurons that they drive, but the goal that our work is directed to is to determine the role of variability in the self-organization of coordinated activity within the group. Stating this in scale-independent terms, the experimental question has to do with the process by which individuals influence the activity of the group, and, in turn, how the group affects the function of the individual. Although our experimental animal may be idiosyncratic in many respects, we believe that the neurointegrative strategies found in it relating to variability and their role in individual and group action may have broad applicability. Moreover, we believe that the problems we have encountered in addressing variability in our experimental animal may also apply to other biological systems.

2.2 Duration and Variability of Biological Responses

Unlike mathematical or physical and chemical systems, such as the Lorenz equations (1963) and the Belousov-Zhabotinskii (Roux et al. 1983) reaction that can be generated indefinitely, biological responses are often short-lived. Moreover, the same behavioral effect may be generated by different movements on different occasions, first, because the external stimuli are probably not the same in those occasions, and, second, because motor patterns encoded within the central nervous system may be inherently variable. Thus, while it is possible to generate thousands of cycles in mathematical attractors, a biological response may be equally well established as an attractor but may last for only several cycles and, thereby, yield only marginally adequate information for characterization of the attractor. For example, bite-ingestion behavior in *Pleurobranchaea* and eating behavior in humans may involve so small a number of repetitive movements of the buccal-oral system that the available analytical tools for chaos may not be applicable, but the behavior may be adaptive and generated by a robust chaotic attractor despite the fact that we may not be able to characterize it as bona fide chaos.

2.3 Multifunctionality, Blending, and Nonstationarity

By multifunctionality we imply, first, the ability to produce different behaviors with the same motor system, and, second, the ability to produce the same behavior in different ways. Blending implies the ability to prodce a behavior that involves a smoothed combination of different behaviors (Bellman 1979; Mpitsos and Cohan 1986 a). Blending can happen progressively; for example, the animal may begin feeding with a particular structure of its motor pattern, and, then as the response progresses, the same effect of feeding may occur but the motor pattern may change into one that has components resembling certin phases of regurgitation. Similarly, when the animal is turned upside down, it may begin to twist its foot in order to right itself, but when inverted and presented food, the foot may attempt to execute two behaviors simultaneously so as to twist in order to right itself, and fold around the food in order to grasp it. In this case, blending happens because of the demands of the stimulus environment, and, therefore, the behavior appears nonstationary because of external forces. However, we have found that deafferented nervous systems themselves can also produce nonstationary motor activity in which an apparently homogeneous, though variable, motor pattern can abruptly or gradually shift into a structurally different pattern (e.g., Mpitsos et al. 1988 c). As in the case of variability in homogeneous responses, the shifts of activity in nonstationary patterns appear to be unpredictable, but further analyses are needed to determine whether they are statistically stationary.

Because of blending, it has not been possible in controlled and blind experiments to identify which behavior the animal was executing solely by examining the electrically recorded motor patterns (Mpitsos and Cohan 1986a). A similar situation occurs in the olfactory EEGs of the rabbit (Freeman and Viana Di Prisco 1986; Freeman and Skarda 1985; Skarda and Freeman 1987). In a remarkable series of experiments, animals were trained to lick to one of two different odors, yet the structure of the EEGs occurring in response to both odors was similar. Having evidence that the animals had learned to distinguish between the two stimuli, Freeman and coworkers reasoned and then demonstrated that the information conveying the different odors must reside within the dynamical aspects of the odor-evoked EEGs, despite the fact that such information could not be distinguished solely from the overall structure of the unprocessed EEG patterns. In our experiments, we recorded the motor patterns from muscles at the same time that the animals repeatedly performed particular behaviors. The electrically recorded motor patterns for similar and different behaviors were quite variable and usually indistinguishable from one another. From this similarity we concluded that the information for the different behaviors probably resided within the dynamics of the motor patterns recorded from the muscles rather than in the appearance of their temporal structure. As we shall discuss below, the process by which this information is generated does not reside exclusively within the nervous system but arises from a greater context resulting from interactions between the animal and the environment.

Because of similar drifts and blends in the activity of single neurons and in the pattern of activity arising from efferent roots, it has not been possible to determine the functional role of individual neurons in establishing motor patterns. This led us to propose:

1. From multifunctionality: That the same neuroanatomic framework consisting to a given set of neurons, their synaptic weights, and background of neuro-modulators can generate different behaviors. That is, connectivity does not determine a behavior but the potentiality of a large set of activities.

To be sure, changes in the neuroanatomic framework, such as selective neuro-modulation and synaptic changes produced by learning, can also lead to the production of different behaviors, but it is necessary to distinguish these possibilities from our proposal which focuses on the emergence of different motor patterns solely through the dynamics occurring within a particular neuroanatomic framework. Harmon's (1964) neuromime studies were the first to show that the same network can produce different patterns of activity. More recent computational studies have confirmed this (e.g., Amit et al. 1983; Kleinfeld and Sompolinsky 1987), and we shall discuss it below regarding chaos in simulation networks. However, because of their uncontrollability, even the simplest biological systems pose difficult experimental problems for studies aimed at proving emergent mul-tifunctionality. Therefore, many questions we wish to address may be answerable only in computer studies that simulate the known characteristics of the biological systems.

2. From context: That knowledge of the functional identity of a given neuron depends on the context of its activity as it takes part in firing with other neurons.

3. From the fluid quality of contextual activity: That the interrelationship of firing in pools of neurons is always shifting because of the inherent variability and chaos in the firing of individual neurons. As it relates to chaos, what comes out of the pool of neurons is deterministic but has long-term unpredictability.

4. From context of function in the activity of sets of synapses: That, given the multifunctional capabilities of the neuronal network, it is conceivable that synaptic changes resulting from training to one adaptive situation may also be used in many different responses such that the notion of context applies not only to groups of neurons but also to groups of synapses. One important implication is that a synaptic change resulting from learning of one behavior does not exclude the possibility that the same synaptic change can code for another learned or un-learned behavior. Whether chaos can act as an informational signal in brain function shall be discussed in the section dealing with simulation networks.

5. From observations of blending, and from findings that motor patterns may not remain stationary nor have a homogeneous flow of variable activity (i.e., the activity can abruptly or smoothly shift from one structure of variable activity into another), it may be useful to consider that:

a. Some attractors may be nonhomogeneous, with each topologically different area representing structurally different neural activity. There may be attrac-tors having long cycles (in the order of minutes) representing the recurrence of blended motor patterns, as well as the shorter cycles (in the order of tens of seconds) representing subcomponents of buccal-oral behavior.

b. The trajectory of the activity may transiently visit different attractors representing different expressions of the same behavior.
c. The different attractors may not be stationary; i.e., there may be a continual, fluid shift in the form of attractors, their position in state space, in their interrelated position to one another, and attractors may blend into one another.
d. A particularly interesting idea to examine is the possibility that metastable chaos may form between the borders of two stable attractors (see discussion of "fuzzy borders" in Decroly 1986).

2.4 The "Right" Attractor

One of the most obvious qualities of animal behavior is that it is prone to error. As we indicated above, a set of neural connections may be able to generate many and variable attractors. Some of these may be adaptive for certain environmental demands, while others may have little adaptability. What determines the appropriate response at any given time, we believe, is not the selection of a particular centrally programmed neural structure or attractor, but rather the process arising from the continual interaction between the animal and the environment. In this respect, the possible fluidity of attractors and the margin left open for error allow for the greatest possible use of the capabilities of a set of connections. When viewed in this way, the concept of attractor is scale-independent: it takes into account a variety of different types of activity that occur within single neurons, among groups of neurons, between areas of the brain, and through the interactions between the animal and the environment.

3 Chaos

3.1 Evidence for Chaos

In order to determine the natural tendency for the central nervous system (CNS) to generate variable activity, we recorded motor patterns in isolated nervous systems so as to remove the effects of sensory inputs (Mpitsos et al. 1988 a, c). Two patterns were examined, one relating to the bite-swallow phase of feeding and one to the active phase of regurgitation. In each case, we examined the activity of a small set of multifunctional neurons in the buccal ganglion and their motoneuron targets in the cerebral ganglion. The buccal-ganglion neurons are essential for coordinating all buccal-oral behaviors and are the major source of the central pattern generator. The recordings were made intracellularly and the spike trains were analyzed using both an interpolation method and an equal-interval sampling method to generate equal-interval time series. The findings of the following analyses provided evidence that the motor patterns were generated by chaotic attractors:

1. The autocorrelation functions of the time series rapidly drop to zero within about four cycles of activity representing the opening and closing of the jaws; i.e.,

the probability for predicting the future evolution of the activity decreases with
time with respect to starting conditions.

2. The value of the principal Lyapunov exponent ($\lambda 1$) was positive for all cells
we examined; the lowest value in bits/s was 0.15 and the highest value was 0.55;
Lyapunov exponents were calculated using the algorithm of Wolf et al. (1985).
In accordance with the autocorrelation functions, this indicates that there was a
gain of information per unit time with respect to initial conditions as the neural
activity evolved (and an equivalent loss in the precision of initial measurements).
The positive value of the exponent indicates that the trajectories exponentially diverge from one another in state space, causing the surface of the attractor to grow
as $2\lambda 1$ in its principal axis; positive growth must be accompanied by folding in
order for the activity to remain bounded.

3. The 1-D maps of successive Poincaré sections through the attractors show
that the trajectories lie in folded sheets that appear to stretch and fold, which is
consistent with the finding of positive Lyapunov exponents. In addition, because
the trajectories lie in sheets rather than at random in the 1-D map, the activity
appears to be deterministic.

4. The correlation dimension of the attractor is low and non-integer (fractal),
and the attractor is embedded in low-dimensional phase space. Values calculated
so far for the correlation dimension range between 1.75 and 2.5 for different
neurons, and the embedding dimensions range from 2 to 6; somewhat higher
values were obtained in calculation of correlation and embedding dimensions of
time series consisting of the unequal spike intervals themselves (unpublished).
The fractal value of the attractors is consistent with chaos (Mandelbrot 1985;
Grassberger and Procaccia 1983). The low-dimensional value of the embedding
space provides some indication that the mechanisms generating the repetitive motor patterns may be simple enough to be analyzable.

3.2 Difficulties in Assessing the Results of Chaos Analysis

Chaos analyses of repetitive activity rely primarily on the number of orbits that
the activity takes through the attractor. A large number of observations may be
useful, but if the number of orbits is too small, high samples do little good. Because the characteristics of biological attractors are usually unknown, experimenters are hard-pressed to determine what constitutes appropriate data lengths
(Albano et al. 1987; and Wolf et al. 1985 discuss some requirements). The short
duration of many biological responses poses difficulties in obtaining sufficient
numbers of orbits. In our experience, homogeneous bite-swallow activity usually
lasts no longer than 10 to 20 cycles; if the attractor is simple enough, there is some
hope of characterizing it. Longer patterns of activity run the risk of nonstationarity. Difficulties arising from nonstationarity have to do with the selection of appropriate homogeneous sections of data for analysis. Thus, while activity having
short periods may appear to yield enough orbits for analysis, other aspects of the
activity may not yield sufficient numbers of orbits for analysis but may represent
an important feature of the same attractor.

The form of the data is also important. Although it is possible to analyze unequally spaced samples (Shaw 1986), more tools are available for data sampled at equal intervals. However, some forms of continuous measures often generate so much data that it may be difficult to analyze in even large computers. For example, a typical bout of 10 feeding cycles in *Pleurobranchaea* may last 200 s. Sampling the intracellular membrane potential of a neuron at a modest rate of 1 kHz generates 200 000 points. Analysis of the correlation dimensions in 10-dimensional state space for this data would require months of CPU time in even the fastest laboratory computer. To avoid long calculation times, the data may be compressed by examining the intervals between action potentials rather then membrane potentials. But in doing so, one now has to handle unequal time intervals. Such data series may be converted to equal-time series (Mpitsos et al. 1988 a, c), but all manipulations in such conversions, and in subsequent handling, run the risk of inducing spurious correlations or removing real ones.

Yet another difficulty arises from the fact that application of the available methods of analysis requires some subjective judgements. For example, there is considerable leeway in the selection of the lag for constructing state space and in selection of slopes in the analysis correlation dimensions, both of which can influence the results (Albano et al. 1987, 1988; Mpitsos et al. 1988 c). Similarly, Lyapunov exponents are sensitive to scaling factors (Wolf et al. 1985; Mpitsos et al. 1988 c). Because of such subjectivity, the exact dimensions of an attractor may be quite difficult to determine. Moreover, given the possible nonstationarities discussed above, the same attractor may not occur twice or similar behavior may be produced by somewhat different attractors. Because nonstationarities are difficult to characterize, their contribution to the measures obtained in a time series may not be obvious to experimenters. Therefore, the validity of similarities or differences in the numbers one obtains from various calculations may be difficult to assess. Because of these problems, evidence of chaos may rely on circumstantial incrimination rather than definitive numerical proof.

3.3 Why is Chaos Important?

Chaos may aid in the generation of organized neural activity. For example, Freeman and Viana Di Prisco (1986), Freeman and Skarda (1985), Skarda and Freeman (1987) found that the rabbit olfactory bulb generates a baseline chaotic EEG that is attributable to chaos. As the animal inhales odorants, the chaotic activity gives way to one of a number of limit cycles that encode specific odors. Our own studies of motor behavior indicate that chaotic attractors themselves may represent the information relating to specific behaviors. Unlike externally added or amplified noise that tends to obscure the neural signal underlying behavior, variability arising from chaos is not separate from this signal, it is the signal. Therefore, by gaining an understanding of the variability arising from chaos, one immediately gains some understanding of the mechanism generating that aspect of the behavior.

Chaos may also have a role in the generation of rapid adaptation to changing environments. Nonvarying signals, such as limit cycles, carry no new information into the future than what they contain at some previous time; activity at one time can be used to predict the evolution at future times: By contrast, the long-term evolution of chaos is not predictable, information at one time does not correlate well with information obtained later on, despite the fact that the activities at the two times are equally deterministic and generated by the same mechanism. Such unpredictability represents a gain of information (e.g., see Abraham and Shaw 1983, vol. 2, p. 107) by which the brain naturally creates new response possibilities.

The sensitivity to initial conditions in chaotic activity may provide a more efficient means for dissipating perturbations than might occur through more predictable mechanisms (Conrad 1986; Mpitsos et al. 1988 a). In chaotic systems, there are as many ways to express a behavior as there are countless trajectories that fill the surface of the attractor that generates it. A small perturbation in the activity at one time leads to a large difference later on, but although the movements of the animal may appear different, their overall effect will be the same; that is, the energy introduced into the system is dissipated through the subsequent deviation in the movements of the animal that produce the same behavior in different ways. By contrast, repetitive activity generated by limit cycles must return asymptotically to the same course of action that occurred before the perturbation. Here, the energy introduced into the system must be dissipated as heat loss to the environment through the tissues of the animal. In either case, sensory feedback and control mechanisms are undoubtedly involved, but the feature that sets chaos apart is that through it the central nervous system itself has a natural tendency to accommodate unexpected inputs.

4 Can the Nervous System Read and Repeat Chaotic Information?

4.1 Simple Networks Learn to Read the Dynamical Features of Chaotic Attractors

A repeating signal embedded in noise has a finite probability of being read (Shannon 1963). In this respect, it is easy to see that limit cycles can provide a readable signal in brain function. Because of their inherent variability and long-term unpredictability, chaotic signals are more difficult to envisage as information sources. Consequently, having obtained evidence that brain function in our experimental animal is chaotic, it was necessary to determine whether such information arising in one part of the nervous system could be read by another.

In order to avoid the complexity of even "simple" neuro/behavioral preparations, we chose to answer this question through computer simulation of simple connectionist networks (Rumelhart et al. 1986; Werbos 1974). Although these networks do not simulate realistic properties of neurons, they nonetheless can be used to answer questions aimed at the flow of information. The network we constructed consisted of one input unit, one output unit, and four hidden (intermediate) units, with each input and output unit being connected to all hidden units (Mpitsos et al. 1988 a). Most studies on such circuits have used binary, nonchaotic signals. Recently, Lapedes and Farber (1988) used chaotic signals in order to determine whether the network could predict the future flow of the input signal. Our study examined the question of whether the network could learn to repeat the 3.60, 3.95 logistic (May 1976) and Rössler (1979) equations, all of which are chaotic.

The results showed that the network can learn to distinguish one chaotic signal from another. Interestingly, the output was always distorted by comparison to the input, and, rather than recognizing specific aspects of the fine details of the input, the network learned the fundamental qualities of the input from which it could generalize to other signals. For example, training the network with random noise proved sufficient for the network to respond immediately to other chaotic attractors as well as it would have following tens of thousands of training trials in the absence of preconditioning with random noise. The one feature that all chaotic attractors have in common is variability. Once the network had learned this quality of the input stimulus through preconditioning, it could immediately characterize other variable inputs. However, the fine parametric details about the input in state space were only approximated by the network, whereas the fundamental aspects of the dynamics that characterized the attractors were reproduced accurately. For example, the Lyapunov exponent $\lambda 1$, which indicates the rate of loss or gain of information in the principal direction of growth in the attractor, was similar in both the input and output attractors. In the same way one might say concerning human perception, that the sensory environment may be perceived first through abstractions of its fundamental qualities. The details of the visual field may be inexact by comparison to the input, and in need of continual updating, even in cases of familiar objects, as, for example, the face of a friend.

4.2   Multifunctionality of Networks

The ability of such networks to learn the fundamental qualities of one chaotic signal and then to generalize from these qualities so as to distinguish rapidly two unlearned chaotic signals was not accompanied by changes in the synaptic weights and thresholds; that is, the network could distinguish several types of signals with the same parametric set of its connections. To be sure, neurophysiological systems can change the type of pattern they produce by using different subsets of the network, different values of the connections, and different neurohumoral modulation (Getting and Dekin 1985; Marder and Hooper 1985). But the network simulations indicate, and agree with our previous proposal (Mpitsos and Cohan 1986a, b), that different types of chaotic motor patterns can also be produced by the same set of connections.

5   Conclusion

In a recent review, Bullock (1984) discussed the need for a revolution in comparative neuroscience. Our own views have arisen from watching animals, and it has seemed to us that a better understanding of behavior requires the merging of two paths that could force such a "quiet revolution." One path holds that self-organized neural output arises from a reciprocal interaction between the animal and the environment. What is not clear to us is whether this dialectic involves simple reductionist cause and effect or whether it leads to higher levels of emergence. Scale-independent concepts such as "attractor" may be useful tools in the study of such a dialectic between animal and environment. What I have attempted to illustrate here is that the description of neural connectivity (including connection weights and neuromodulators) does not determine a behavior, but rather the possibility of many behaviors, some representing adaptive responses and some representing error or nonsense. What determines the functional "neurocircuit" is the relationship of the animal to the environment. Without the dynamic between animal and environment, one can examine such a network in detail and not know what it can do. One is reminded here of the analogy drawn by Gould and Lewontin (1979) between the capability of the brain and paintings placed in the spandrels arising from intersection of cathedral arches: structure provides the bounds, but what becomes expressed through structure is determined by external input.

The second path holds that variability is an integral and inseparable part of such neurointegrative mechanisms, not just an expression of sloppy biological function. Chaos is interesting because of its deterministic character, but nondeterministic processes such as thermal or random noise may also be important (Kirkpatrick et al. 1983; Jeffrey and Rosner 1986; Burton and Mpitsos 1988a, b). We have discussed some of the major difficulties in assessing whether chaos occurs in behavioral processes. Given the weight of some of these problems, it may not be possible to prove beyond circumstantial evidence whether chaos actually exists in a particular behavior. But even as a heuristic tool, chaos may provide access to new formulations about brain function because unlike added or amplified noise, it does not have a separate existence from the neural code underlying behavior: to the extent that chaotic variation exists in motor processes, it is the code.

6   Summary

Our studies of variability of buccal-oral behaviors in the sea slug *Pleurobrancheae californica,* and simulation of network activity, has led us to propose that it may not be possible to assign specific functional identities to neurons, their synapses, and even to groups of neurons, first, because they can produce qualitatively different motor patterns (they are multifunctional), and second, because the interrelated firing of neurons can change even during the expression of a given motor pattern. While reflexes undoubtedly occur in neural function, the emergent properties that arise from variability in parallel processing may go beyond the reflex in the same way that quantum physics superceded classical physics. Whereas classical reflex neurophysiology may propose that a learned behavior is ascribable to a particular synaptic change, the newer perspective cautions that the same change may be used in other learned and even unlearned behaviors, and that variability in the responses of individual neurons, and in the interrelated responses of groups of them, plays a central role in generating behaviors. Therefore, we define three integrative levels: anatomical, dynamical, and self-organizing, representing, respectively, the structural framework, the temporal characteristic of linear and nonlinear processes, and emergent qualities. Self-organization arises from the first two, but, through the quality of variability, it also supercedes them.

Although there may be many sources of behavioral variability, as may arise from random noise in neural processing or from perturbations produced by sensory inputs, our analyses of motor patterns in deafferented nervous systems indicate that chaos in the responses of individual neurons may be one contributing factor. The frequency of firing in these neurons was examined by means of state-space analyzes and was found to be unpredictable over the long-term, but, nonetheless, to be generated by deterministic mechanisms: 1) Autocorrelation functions of spike-frequency time series rapidly decreased, indicating that the activity was sensitive to initial conditions. 2) Accordingly, phase portraits exhibited state-space trajectories that diverged and mixed, and 3) the value of the principle Lyapunov exponent was positive, providing a quantitative measure of the exponential divergence of nearby trajectories which would cause the surface of the phase portrait to stretch. 4) Poincaré sections taken through the phase portraits provided graphic evidence of folding that should accompany stretching. 5) From calculation of integral autocorrelation functions, phase portraits were found to have fractal correlation dimensions (ranging from about 1.75 to 2.5) embedded in low-dimensional state space (from 2 to 6 dimensions). Together, these findings indicate that the activity was generated by chaotic attractors, and the low-dimensionality of the activity suggests that the underlying processes may be simple enough for further experimental study.

Whatever the source of variability, the aspect of animal behavior, and probably of other biological systems, that most taxes the application of dynamical theories is that it can be nonstationary. By nonstationary, we imply the ability of patterns of activity to change abruptly or blend gradually into other patterns that appear qualitatively different from the preceding though variable activity, and the pattern overall may not be statistically stationary. Although such changes may represent shifts into different behaviors, they can also occur during the execution of the same ongoing behavior. In addition, significant differences may occur between repetitions of either short- or long-duration behaviors. Such nonstationarities may occur over the long-term, but they also may occur within several cycles in behaviors that involve many repetitions. Because of graded blending, it is often difficult to determine whether the pattern changes constitute nonstationarities or are simply part of the long-term variability in a stationary pattern. Finally, even when robust, stationary attractors generate adaptive biological responses, the duration of the responses is often too short for application of the available mathematical tools. Together, these aspects of biological systems compromise the applicability of extant dynamical theory which has been developed primarily through well-defined mathematical and physical systems. Consequently, we believe that numbers arising from application of the available chaos theory must be viewed cautiously, more as indicators of the possible qualities rather than as absolute measurements of particular dimensions or characteristics of the underlying dynamics.

Given that adaptive behaviors may be generated by chaotic attractors, the unpredictability of such attractors required us to ask whether activity arising in one part of the nervous system could be read and reproduced by another. Because there are no neuro/behavioral systems that are sufficiently simple and controllable for inquiring into this question quantitatively, we studied the informational aspects of the problem in computer simulations of a simple connectionist network consisting of one input, one output, and four hidden units. As determined by quantitative comparisons of the input chaotic signal with the output of the network, recognition and repetition of such signals appear to be accompanied by distortion. What is learned or remembered has more to do with fundamental qualities in the dynamics of the input signal, such as the rate of loss of information or the structure of folding in the attractor surface, rather than the specifics of its parametric details such as the value of the input at each successive interval. It seems that after learning some fundamental quality of a signal, the same underlying network of connection weights and thresholds can repeat other, previously unlearned signals as well as if it had learned them first; i.e., the ability to respond to different signals arises from generalization through the characteristics that are shared by these signals. For example, after being trained to a random signal, the network can recognize and repeat different chaotic signals because they share in the quality of appearing random, but whether generalized or learned, the distortion in the output is the same.

Thus, our findings on physiological and computational systems indicate that connectivity does not determine a behavior, but rather the potentiality of many behaviors arising as discrete and separate from one another or as blends, some of which may be adaptive and some may represent errors. We have proposed that variability arising from chaos, random noise, and feedback from the environment are essential parts of such self-organizing processes. The aspect of chaos that makes it particularly interesting is that through its tendency to gain information with respect to initial conditions i.e, the nervous system itself has a natural capacity to generate new informational space for adaptation to an unpredictable environment.

Acknowledgments. I thank the Air Force Office of Scientific Research for support (grant 86-0076), W. Ross Adey, Robert M. Burton, Theodore H. Bullock for critical reading of a previous version of this manuscript, and Lavern Weber for making space available for me at the Hatfield Marine Science Center.

References

Abraham RH, Shaw CD (1983) Dynamics. The geometry of behavior, vols 1–3. Aerial, Santa Cruz
Adey RW (1972) Organization of brain tissue: is the brain a noisy processor? Int J Neurosci 3:271–284
Aihara K, Matsumoto G (1986) Chaotic oscillations and bifurcations in squid giant axons. In: Holden AV (ed) Chaos. Princeton University Press, Princeton, pp 257–269
Albano AM, Mees AI, de Guzman GC, Rapp PE (1987) Data requirements for reliable estimation of correlation dimensions. In: Degn H, Holden AV, Olsen LF (eds) (1986) Chaos in biological systems. Plenum, New York

Albano AM, Muench J, Schwartz C, Mees AI, Rapp PE (1988) Singular-value decomposition and the Grassberger-Procaccia algorithm. Physical Rev 38A:3017–3026

Amit DJ, Gutfreund H, Sompolinsky H (1985) Storing infinite numbers of patterns in a spin-glass model of neural networks. Phys Rev Lett 55:1530–1533

Babloyantz A, Destexhe A (1986) Low dimensional chaos in an instance of epilepsy. Proc Natl Acad Sci USA 83:3513–3517

Babloyantz A, Salazar JM, Nicolis C (1985) Evidence of chaotic dynamics of brain activity during the sleep cycle. Phys Lett 111A:152–155

Bellman KL (1979) The conflict behavior of lizard, *Sceloporus occidentalis*, and its implication for the organization of motor behavior. PhD dissertation, University of California, San Diego

Bullock TH (1984) Comparative neuroscience holds promise for quiet revolution. Science 222:473–478

Burton RM, Mpitsos GJ (1988 a) A constant-noise mechanism for enhancing learning in neural networks and its comparison to simulated annealing. Soc Neurosci Abstr 14:104–114

Burton RM, Mpitsos GJ (1988 b) Noise enhanced learning in neural networks: comparison to simulated annealing, and applicability to biological systems. (To be published)

Chay TR (1985) Chaos in a three-variable model of an excitable cell. Physica 16D:233–242

Cohan CS, Mpitsos GJ (1983 a) The generation of rhythmic activity in a distributed motor system. J Exp Biol 102:25–42

Cohan CS, Mpitsos GJ (1983 b) Selective recruitment of interganglionic interneurons during different motor patterns in *Pleurobranchaea*. J Exp Biol 102:43–58

Conrad M (1986) What is the use of chaos? In: Holden AV (ed) Chaos. Princeton University, Press, Princeton, pp 3–14

Decroly O (1986) Interplay between two periodic enzyme reactions as a source for complex oscillatory behavior. In: Degn H, Holden AV, Olsen LF (eds) Chaos in biological systems. Plenum, New York, pp 49–58

Degn H, Holden AV, Olsen LF (eds) (1986) Chaos in biological systems. Plenum, New York

Freeman WJ, Skarda CA (1985) A perspective on brain theory: nonlinear dynamics of neural masses. Brain Res Rev 10:147–175

Freeman WJ, Viana Di Prisco GV (1986) EEG spatial pattern differences with discriminated odors manifest chaotic and limit cycle attractors in olfactory bulb of rabbits. In: Palm G, Aertsen A (eds) Brain theory. Proceedings the first Trieste meeting on brain theory, 1–4 October, 1984. Springer, Berlin Heidelberg New York

Getting PA, Dekin MS (1985) *Tritonia* swimming: a model system for integration within rhythmic motor systems. In: Selverston AI (ed) Model neural networks and behavior. Plenum, New York, pp 3–20

Gleick J (1987) Chaos. The making of a new science. Viking-Penguin, New York

Gould SJ, Lewontin RC (1979) Spandrels of San-Marco and the panglossian paradigm: a critique of the adaptationist program. Proc R Soc Lond [Biol] 205–581–588

Grassberger P, Procaccia I (1983) Characterization of strange attractors. Phys Rev Lett 50:346–349

Grossberg S (1980) Studies of mind and brain. Reidel, Boston

Harrnon LD (1964) Neuromimes: action of a reciprocally inhibitory pair. Science 146:1323–1325

Hayashi H, Ishizuka S (1986) Chaos in molluscan neuron. In: Degn H, Holden AV, Olsen LF (eds) Chaos in biological systems. Plenum, New York, pp 157–166

Holden AV (ed) (1985) Chaos. Princeton University Press, Princeton

Hopfield JJ (1982) Neural networks and physical systems with emergent computational abilities. Proc Natl Acad Sci USA 79:2554–2558

Jeffrey W, Rosner R (1986) Optimization algorithms – simulated annealing and neural network processing. Astrophys J 310:473–481

John ER (1972) Switchboard versus statistical theories of learning and memory. Science 177:850–864

Kelso JAS, Scholz JP, Schöner G (1986) Nonequilibrium phase transitions in coordinated biological motion: critical fluctuations. Phys Lett 18A:279–284

Kirkpatrick S, Gelatt CD, Vecchi MP (1983) Optimization by simulated annealing. Science 220:671–680

Kleinfeld D, Sompolinsky H (1988) Associative neural network model for the generation of temporal patterns: theory and application to central pattern generators. Biophysical J 54:1039–1051

Labos E (1986) Chaos and neural networks. In: Degn H, Holden A V, Olsen LF (eds) Chaos in biological systems. Plenum, New York, pp 195–206

Lapedes A, Farber R (1987) Nonlinear signal processing using neural networks: predictive and system modeling. Los Alamos Laboratory Technical Report, LA-UR-87-2662

Lorenz EN (1963) Deterministic non-periodic flows. J Atmos Sci 20:130–141

Mandelbrot BB (1985) The fractal geometry of nature. Freeman, New York

Marder E, Hooper SL (1985) Neurotransmitter modulation of the stomatogastric ganglion of decapod crustaceans. In: Selverston A I (ed) Model neural networks and behavior. Plenum, New York, pp 319–338

May RM (1976) Simple mathematical models with complicated dynamics. Nature 261:459–467

McClellan AD (1982a) Movements and motor patterns of the buccal mass of *Pleurobranchaea* during feeding, regurgitation, and rejection. J Exp Biol 98:195–211

McClellan AD (1982b) Re-examination of presumed feeding motor activity in the isolated nervous system of *Pleurobranchaea*. J Exp Biol 98:212–228

Mpitsos GJ, Cohan CS (1986a) Comparison of differential Pavlovian conditioning in whole animals and physiological preparations of *Pleurobranchaea:* implications of motor pattern variability. J Neurobiol 7:499–516

Mpitsos GJ, Cohan CS (1986b) Convergence in a distributed motor system: parallel processing and self-organization. J Neurobiol 7:517–545

Mpitsos GJ, Collins SD, McClellan AD (1978) Learning: a model system for physiological studies. Science 199:497–506

Mpitsos GJ, Burton RM, Creech HC, Soinila SO (1988a) Evidence for chaos in spike trains of neurons that generate rhythmic motor patterns. Brain Res Bull 21:529–538

Mpitsos GJ, Burton RM, Creech HC (1988b) Connectionist networks learn to transmit chaos. Brain Res Bull 21:539–546

Mpitsos GJ, Creech HC, Cohan CS, Mendelson M (1988c) Variability and chaos: neurointegrative principles in self-organization of motor patterns. In: Kelso JAS, Mandell AJ, Shlesinger MF (eds) Dynamic patterns in complex systems. World Scientific, Singapore, pp 162–190

Pellionisz A, Llinas R (1983) Space-time representation in the brain: the cerebellum as a predictive space-time metric tensor. Neuroscience 7:2949

Rapp PE, Zimmerman ID, Albano AM, Deguzman GC, Greenbaum NN (1985) Dynamics of spontaneous neural activity in the simian motor cortex: the dimension of chaotic neurons. Phys Lett 10A:335–338

Rössler OE (1979) An equation for continuous chaos. Phys Lett 57A:397–398

Roux J-C, Simoyi RH, Swinney HL (1983) Observation of a strange attractor. Physica 8D:257–266

Rumelhart DE, McClelland JL, PDP Group (eds) (1986) Parallel distributed processing: explorations in the microstructure of cognition, vol 1. Foundations. MIT Press, Cambridge

Sejnowski TP, Rosenberg CR (1987) Parallel networks that learn to pronounce English text. Complex Syst 1:145–168

Selverston AI (1980) Are central pattern generators understandable? Behav Brain Sci 3:335–371

Shannon CE (1963) The mathematical theory of communication. In: Shannon CE, Weaver W (eds) The mathematical theory of communication. University of Illinois Press, Urbana (Reprinted from Bell Syst Tech J 1948, July and October)

Shaw R (1986) The dripping faucet. Aerial, Santa Cruz

Skarda CA, Freeman WJ (1987) How brains make chaos in order to make sense of the world. Behav Brain Sci 10:161–195

Sperry RW (1981) Changing priorities. Ann Rev Neurosci 4:1–15

Werbos PJ (1974) Beyond regression. New tools for prediction and analysis in the behavioral Sciences. PhD thesis. Harvard University, Cambridge

Wolf A, Swift JB, Swinney HL, Vastano JA (1985) Determining Lyapunov exponents from a time series. Physica 16D:285–317